Standard Operating Procedures and Guidelines

Standard Operating Procedures and Guidelines

John Lee Cook, Jr.

Disclaimer

The recommendations, advice, descriptions, and methods in this book are presented solely for educational purposes. The author and publisher assume no liability whatsoever for any loss or damage that results from the use of any of the material in this book. Use of the material in this book is solely at the risk of the user.

Published by Fire Engineering Books & Videos
A Division of PennWell Publishing Company
Park 80 West, Plaza 2
Saddle Brook, NJ 07663
United States of America

James J. Bacon, editor
Book design by Patricia Rasch
Cover design by Steve Hetzel

1 2 3 4 5 6 7 8 9 10

Printed in the United States of America

Library of Congress Cataloging-in-Publication Data

Cook, John Lee, Jr., 1950-
Standard operating procedures and guidelines / John Lee Cook, Jr.
p. cm.
ISBN 0-912212-69-1 (softcover)
1. Fire departments--Rules and practice. 2. Fire departments--Management.
I. Title.
TH9158.C66 1998
363.37′068—dc21

98-43459
CIP

About the Author

John Lee Cook, Jr., entered the fire service in 1969 when he was voted in as a member of the Huntsville (TX) Volunteer Fire Department. Since that time, he has served as a career firefighter, training officer, and college instructor. He served as fire chief and emergency management coordinator in Conroe, Texas, from 1981 to 1987, and in Denton, Texas, from 1987 to 1995. Cook served as the director of the Department of Fire and Rescue Services in Loudoun County, Virginia, from 1996 to 1998.

Cook has a bachelor's degree in business administration from Sam Houston State University and a master's in public administration from Southwest Texas State University. He is a charter member of the Fire Service Section of the National Fire Protection Association and served on the Technical Committee for Fire Service Training. He is a member of the International Association of Fire Chiefs, the State Fire Chiefs Assocation of Virginia, and the National Fire Sprinkler Assocation. He is an associate member of the Institution of Fire Engineers.

Cook has published numerous articles in fire service periodicals and is frequently invited to speak to various fire service gatherings. He and his wife, Elizabeth, reside in Leesburg, Virginia.

This book is dedicated to the brave men and women with whom I have had the pleasure of serving during my career in the fire service. You have taught me much and you have been so gracious as to allow me to inflict my theories of management on you. Thank you and may God bless and protect you.

Table of Contents

Introduction 1

Section 100
Rules and Regulations 11

100.01, Administration, *13*
100.02, Definitions, *14*
100.03, Acting in a Higher Classification, *16*
100.04, Equal Employment Opportunity, *17*
100.05, Disciplinary Actions, *17*
100.06, Grievance Procedure, *17*
100.07, Leave and Vacations, *18*
100.08, Media Relations, *19*
100.09, Resignations, *19*
100.10, Promotions, *19*
100.11, Seniority List, *20*
100.12, Shift Swaps, *21*
100.13, Temporary Appointments, *21*
100.14, Transfers, *22*
100.15, Workers' Compensation/Injury Leave, *22*
101.01, Fire Chief—Job Description, *23*
101.02, Deputy Chief of Emergency Services—Job Description, *23*
101.03, Deputy Chief of Support Services—Job Description, *24*
101.04, Deputy Chief of Technical Services—Job Description, *24*
101.05, Captain—Job Description, *25*
101.06, Driver—Job Description, *26*
101.07, Firefighter/EMT (Firefighter/Paramedic)—Job Description, *26*
101.08, Assistant Fire Marshal—Job Description, *27*
101.09, Investigator—Job Description, *27*
101.10, Inspector—Job Description, *28*
101.11, Training Captain—Job Description, *28*
101.12, Training Officer—Job Description, *29*
101.13, Communications Supervisor—Job Description, *30*
101.14, Communications Shift Supervisor—Job Description, *30*
101.15, Dispatcher—Job Description, *31*
101.16, Emergency Management Coordinator—Job Description, *31*

101.17, Administrative Manager—Job Description, *32*
101.18, Secretary—Job Description, *32*
101.19, Volunteer Coordinator—Job Description, *33*
102.01, Code of Conduct, *33*
102.02, Obedience to Orders, *34*
102.03, Professional Relations, *35*
102.04, Personal Appearance, *35*
102.05, Physical and Mental Fitness, *36*
102.06, Recall to Duty, *36*
102.07, Station Duty, *36*
102.08, Uniforms, *37*
102.09, Vehicles and Equipment, *39*
102.10, Visitors at Stations, *40*
103.01, Controlled Substances, *41*
103.02, Inappropriate Behavior, *41*
103.03, Outside Employment, *42*
103.04, Sexual Harassment, *42*
103.05, Use of Tobacco Products, *43*
103.06, Employees Serving as Volunteers, *43*

Section 200

General Administration 45

200.01, Station Supplies, *47*
200.02, On-Duty Meals, *48*
200.03, Lost/Destroyed Equipment, *49*
200.04, Fire Department Library, *50*
200.05, Daily Work Schedule, *51*
201.01, Minimum Staffing, *53*
201.02, Collateral Assignments, *56*
201.03, Complaints Against Employees, *58*
201.04, Evaluation of Sworn Personnel, *61*
202.01, Shift Training Officers, *62*
203.01, Training/Travel Requests, *63*
203.02, Compensation/Reimbursement for Travel, *63*
203.03, Records and Forms, *65*
203.04, Incident Reports, *66*
203.05, Patient Treatment Forms, *68*

Section 300

Hazardous Materials 71

300.01, Program Management, *73*
301.01, Emergency Response, *74*
301.02, Containment and Cleanup, *76*
301.03, Decontamination Procedures, *77*
301.04, Personnel Safety, *78*
301.05, Contaminated Protective Clothing, *80*

Section 400
Occupational Safety and Health 81

400.01, Occupational Safety and Health Program, *83*
400.02, Fire Department Safety Officer, *84*
400.03, Occupational Safety and Health Committee, *84*
400.04, Records, *85*
401.01, Operating Hydraulic-Powered Rescue Tools, *86*
401.02, Operating Power Saws, *88*
402.01, Training, *90*
402.02, SCBA Training, *91*
402.03, SABA Training, *92*
403.01, Drivers of Vehicles, *93*
403.02, Persons Riding in Motorized Vehicles, *94*
403.03, Emergency Response, *95*
403.04, Vehicle Accident Review Board, *97*
404.01, Fire Station Safety, *98*
405.01, Protective Clothing and Equipment, *99*
405.02, Hearing Conservation, *102*
405.03, Flashover/Backdraft Report, *103*
405.04, Rehabilitation, *103*
405.05, Operating at Emergency Incidents, *108*
405.06, Accountability, *110*
406.01, Fitness for Duty, *113*
406.02, Permanent Medical File, *115*
406.03, Exposure Reporting, *116*

Section 500
Maintenance 119

500.01, Repair Requests, *121*
501.01, Apparatus and Motorized Vehicles, *122*
501.02, Equipment Assigned to an Apparatus, *125*
501.03, Declaring a Vehicle Unsafe to Operate, *128*
501.04, Marking Vehicles and Apparatus, *130*
502.01, Marking and Inventorying Equipment, *132*
503.01, Fire Hose, *135*
503.02, Fire Hose Testing, *136*
503.03, Fire Hose Records, *138*
504.01, Self-Contained Breathing Apparatus, *139*
505.01, Ground Ladders, *142*

Section 600
Emergency Operations 145

600.01, Glossary of Terms, *147*
600.02, Incident Command System, *149*
600.03, Minimum Company Standards, *150*
600.04, Postmortems, *163*
600.05, Tactical Guidelines, *164*
600.06, Tactical Surveys, *167*
601.01, High-Rise Buildings, *170*

601.02, High-Rise Operations, *171*
602.01, Water Rescue, *172*
603.01, Emergency Vehicle Placement, *174*
603.02, Personnel Deployment on the Fireground, *176*
603.03, Fire Streams, *180*
603.04, Hose Loads, *182*
603.05, Water Supplies, *184*
603.06, Ventilation, *187*
604.01, Support of Automatic Sprinkler Systems, *188*
604.02, Standpipe Operations, *189*
605.01, Airport Response, *190*
605.02, Motor Vehicle Fires, *192*
606.01, Progressive Hoselays, *195*
606.02, Wildland Fires, *196*

Section 700

Emergency Medical Services **201**

700.01, General Patient Care, *203*
700.02, Deceased Persons, *204*
700.03, Terminally Ill Patients, *205*
700.04, Mass-Casualty Incidents, *206*
701.01, Patient Transport, *208*
701.02, Emergency Transfers, *209*
701.03, Transport by Helicopter, *210*
702.01, Infection Control Program, *212*
702.02, Decontamination, *214*
702.03, Protective Clothing, *216*

Section 800

Communications **219**

800.01, Glossary, *221*
801.01, Radio Procedures, *227*
801.02, Radio Numbers, *231*
802.01, Radio Maintenance, *234*
803.01, Dispatching Alarms, *234*
803.02, Daily Radio Tests, *240*
803.03, Dispatching the Emergency Management Coordinator, *241*
803.04, Paging, *242*
804.01, Enhanced 911, *243*
804.02, Emergency Ring-Down Circuits, *244*
804.03, Cellular Telephones, *244*
804.04, Department Telephone System, *245*

Section 900

Fire Prevention **247**

900.01, Weapons, *249*
900.02, The Use of Deadly Force, *251*
901.01, Company Inspections, *252*
901.02, Determination of Cause and Origin, *255*

Appendix 257

1. Employee Leave Request Form, *259*
2. Daily Activity Report, *260*
3. Driver's Daily Apparatus Checklist, *261*
4. Station Supply Requisition Form, *262*
5. Report of Equipment Lost or Destroyed, *263*
6. Library Book Check-Out Log, *264*
7. Audiovisual Materials Check-Out Form, *265*
8. Complaint Form, *266*
9. Request for Travel/Training Form, *267*
10. Annual Employee Performance Review Form, *268*
11. Basic Incident Report, *272*
12. Basic EMS Report, *274*
13. Basic Casualty Report, *276*
14. EMS Charge Sheet, *278*
15. Hazardous Materials Exposure Form, *279*
16. Notification of Accident Report, *280*
17. On-the-Job Injury Supervisor's Report, *282*
18. Supplemental Report of Injury, *284*
19. Designation of Duty Status Form, *286*
20. Authorization to Ride an Apparatus Release, *287*
21. Inspection Report Form, *288*
22. Flashover/Backdraft Report, *290*
23. Rehab Log, *291*
24. Rehab Unit Checklist, *292*
25. Equipment Service Request Form, *293*
26. Small-Vehicle Weekly Checklist, *294*
27. Annual Fire Pump Service Test Form, *295*
28. Preventive Maintenance Worksheet, *296*
29. Red Tag Out-of-Service Card, *298*
30. Master Hose Record Card, *299*
31. Monthly SCBA Inspection Form, *300*
32. Ground Ladder Record Form, *301*
33. Tactical Worksheet, *302*
34. Minimum Company Standards Evaluation Form, *304*
35. Tactical Survey Form, *307*
36. Directive to Physicians, *310*
37. Triage Tag, *312*
38. Employee Immunization Refusal Form, *316*
39. Alarm and Fire Record Card, *317*
40. Reinspection Fee Schedule, *318*

Introduction

Leaders define themselves by what they enforce.
—Alan Brunacini

In an era in which managerial authoritarianism is being replaced with participative management and employee empowerment, formal rules and guidelines have fallen into disrepute. Even so, participative management doesn't connote *permissive* management. Participative management can only work on top of an existing organizational structure. Every organization needs a set of rules to govern its members. Without rules, chaos reigns. Moreover, rules communicate as well as control—they let an organization's members know what is expected of them. An organization with rules is an organization in which there are no surprises.

While not all-inclusive, this book contains a collection of sample standard operating procedures for a wide range of fire department activities, including emergency medical services. These sample procedures are drawn from my own experiences as well as those of a number of fire service organizations and agencies, large and small. This collection of SOPs is meant to serve as a template on which a fire department can base its own manual or procedures. One size does not fit all. It isn't essential to use all of the sample procedures in this book. You won't compromise your department's manual by omitting procedures that aren't applicable to your organization. Each one that you adopt should be specific to the size, mission, and scope of your department. Likewise, just because a subject hasn't been addressed in this book doesn't mean your department doesn't need procedures that address it. Your department's mission and the services it provides dictate which procedures to include in your SOPs.

The terms that follow have meanings germane to management. You should become familiar with them.

General order—A written, numbered directive that changes a department's rules, regulations, policies, or standard operating procedures. In most fire departments, general orders (as well as SOPs) are issued by the fire chief, although in some larger departments a senior officer may issue them.

Guideline—A statement of policy by a person or group having authority over an activity.

Policy—A plan or course of action designed to influence and determine decisions and actions. A policy's purpose is to ensure consistency of action and provide guidelines for a given activity.[1]

Procedure—While policy guides decision making, a procedure is a more detailed guide to action. A procedure details the steps to be followed, and in what order, in carrying out an organizational policy related to a specific recurring problem or situation.[2]

For example, a department may have a policy requiring that command at all working structure fires be established in accordance with the department's incident command system. The procedures supporting that policy would describe the components of the incident command system and how to use them during an incident.

Regulation—A general term referring to a principle, rule, or law that governs behavior.

Standard operating procedure—A written procedure aimed at standardizing general activities.

Rules and Procedures—How and Why They Should Be Promulgated

Henri Fayol, the first author to identify management as a process, recognized the importance of control in managing an organization. Control is necessary for verifying whether the actions of the members are in accord with the organization's adopted plans and underlying principles. Control involves monitoring for system weaknesses, human errors, and deviations from the assigned path. It allows discrepancies to be rectified and prevents their occurrence. Control touches on all aspects of an organization: people, actions, objects, and timing.[3]

Rules are part of control and an essential part of management. In any organization, control is necessary to minimize risk and ensure predictable outcomes during standard operations. Consistency of outcome and risk minimization are essential aspects of safe and efficient fireground operations—fire suppression, after all, is a matter of life and death. When outcomes matter the most, three basic rules of risk management apply:

—Don't risk more than you can afford to lose.
—Don't risk a lot for a little.
—Consider the odds.[4]

Remember that established procedures allow managers and employees to make decisions more rapidly and with the confidence that the results from one operation to the next will be uniform, thus enhancing fire department efficiency. Rules give employees a greater sense of security. When there are rules, employees know what treatment they can expect in a given situation. Leaders are able to act with a greater degree of confidence in resolving problems with SOPs in place, since they have an objective basis on which to make and defend their decisions.[5] When SOPs are the source of authority, a manager's decision is less likely to be questioned.

Having said that, keep in mind the following principles with respect to the rules and procedures your department promulgates.

Your SOPs should be appropriate for your department. Standard operating procedures should take into account the conditions in your locale and the capabilities and limitations of your department. At the same time, you do not need to reinvent the wheel. If a particular SOP has worked well for another department, it will likely work for yours. Get on the telephone with emergency service managers in neighboring communities and ask to see their SOPs, noting the differences between their organizations and your own. Your SOPs must be a good fit for your department. Still, we are often not as unique as we would like to believe. If it works on the West Coast, it will probably work on the East Coast. It may be especially useful to try to borrow from large departments, which typically have far more collective experience on a greater range of subjects than small departments.

Standard operating procedures must be in writing. Section 6-1.2 of NFPA 1500, *Standard on Fire Department Occupational Safety and Health Program,* requires fire departments to adopt an incident management system that meets the criteria of NFPA 1561, *Standard on Fire Department Incident Management System,* and to establish the system with written SOPs that apply to all members involved in emergency operations. It follows that your other operational procedures should be committed to writing. Unwritten directives are difficult to learn, remember, and apply.[6] Committing SOPs to writing also allows members to focus on critical rather than routine decisions.

The adoption and use of written SOPs involves a major change from an unstructured organization to a more structured one. Written SOPs make a department more professional. They eliminate the game of trying to guess what will happen next. Having written procedures helps reduce freelancing by individual members. Written SOPs also make a department more cohesive.[7]

There is, of course, an unlimited litany of excuses for not developing written SOPs or for not following them if you do have them. If your department relies entirely on volunteers, you may be tempted to believe that you cannot tell volunteers what to do. If your department is small, you may think that SOPs are something only the big-city departments want or need. Or maybe your department has had SOPs in the past that no one followed. None of these arguments are valid. Fires do not discriminate based on employment status, geographical location, or the degree of commitment on the part of management. Fires do, however, discriminate between those who are prepared and those who are not.

Standard operating procedures must be followed. Written SOPs are only effective if they are used. An SOP that management doesn't enforce isn't a true SOP. Enforcement should be educational, providing the opportunity for positive rather than negative reinforcement. One way of reinforcing and institutionalizing SOPs is to include them as source material for promotional examinations. Knowing and following SOPs helps individuals and groups develop a set of good work habits.

Following SOPs can also protect against arbitrary action. Each of us is sensitive to differences, no matter how slight, in how we are treated compared with other persons. Following standard procedures with respect to every department member helps eliminate actual or perceived differences in treatment and the problems they can create. Not following SOPs can have serious legal ramifications, as discussed in greater detail below. If an SOP is impossible to enforce, something has to change—the organization, its leadership, or the policy.

Policies should aid rather than hinder decision making. Standardizing operating procedures and putting them in writing should simplify rather than complicate matters. Policies must not discourage an officer's use of his own judgment when searching for a solution to a given problem. Nor should the officer use the department's policies as an excuse for not taking action or approving a request. Policies must be dynamic. They must allow for change depending on the circumstances. Remember, a policy is a means to an end, not an end in itself or an excuse for failure to take needed action.

Standard operating procedures should be official. An official SOP is one that is sanctioned by the department, as demonstrated by a written general order from the fire chief establishing or enacting it. An SOP becomes official only after every member has been notified of it in writing. The fire station lawyers would argue that, if it isn't in writing, it isn't official and therefore may not necessarily apply in a given situation. It is easier to sanction or discipline a member of the organization for violating a written rule. It is hard to demonstrate that an unwritten rule even exists. If it was that important, the argument goes, why wasn't it written down?

Litigation Issues

In our litigious age, some people conclude that the mere presence of written procedures invites litigation and therefore should be avoided at all costs. Unfortunately, in our line of work, litigation is like risk—it is unavoidable. The presence of a written SOP, ideally based on a national standard, indicates that your department is aware of the importance of that activity. In court, you department's lawyer will be able to argue that the written SOP indicates that members know what is expected of them and are instructed on how to perform a certain activity. The lawyer will also be able to argue that members performed according to those expections the same way that the members in other departments perform on a routine basis throughout the country.

Basing SOPs on nationally recognized standards and applicable state or local laws, codes, and regulations will make things easier for your department should litigation occur. The broad base of acceptance a national standard enjoys, plus the length and complexity of the process involved in its development, will work in your department's favor.[8] The prevailing expectation is that a knowledgeable, prudent person would follow the directives of an applicable national standard. Therefore, anyone attempting to find the department liable will have to focus on negligent action rather than attacking the contents of the department's SOP.

Members must know the department's SOPs. At issue in litigation will be whether an SOP was followed. One of the first things the opposing attorney will do in preparing a case against you is obtain a copy of the department's SOP manual. When the matter goes to court, you can bet that he will be conversant with the department's SOPs. Hopefully, you will be equally familiar with your own manual.

No department is immune from liability issues. To give a personal example, a department in which I served as fire chief was forced to pay damages to an injured worker because the incident commander and several company officers had failed to follow the department's SOPs. The incident that gave rise to the litigation was the department's response to a fire at the site of a building undergoing demolition, where a trapped worker was injured. The court found that, had the department followed its own SOPs, the worker wouldn't have been allowed to remain inside the building and probably wouldn't have been injured. My point is simply that members should know the department's procedures and, when a procedure applies, should follow it.

What your procedures are called, whether they are in writing, and how they are written may all become relevant in the event of litigation. There is, for instance, some debate over whether directives should be called "procedures" or "guidelines." Many people interpret "procedure" to mean "follow or else," whereas "guideline" implies some degree of flexibility. The assumption is that a department's liability might be reduced or avoided altogether in the event a department has "guidelines" rather than "procedures." Calling your directives one thing over another isn't a foolproof way of avoiding lawsuits, nor is it likely to make much difference in the outcome of litigation.

Of greater importance is whether a given directive is written in mandatory or permissive terms. The words *shall* and *will* within a document may become extremely important in the event your department's activities come under scrutiny. *Shall* and *will* are imperative words, and their use may not only unnecessarily restrict the actions of the incident commander but also give your courtroom opponent an advantage. *Should* or *may,* on the other hand, indicate that an action is a recommendation—advised but not required. This provides the user of the SOP with greater flexibility, allowing for deviation from the SOP's directives when appropriate. It might even be a good idea, where appropriate, to include in the SOPs a statement to the effect that a commanding officer is allowed to deviate from the procedures when unusual circumstances require it.

Developing and Implementing SOPs

Getting started. Deciding to develop SOPs for your department is like deciding to start jogging—the hardest part is putting on your shoes. Once you are ready to go, the actual activity, while not painless, isn't as difficult as you anticipated. And in the long run, the results are beneficial. There are, of course, practical elements involved. Perhaps the most important tool in writing effective SOPs is a good copying machine. A copier will allow you to duplicate other departments' SOPs so you can read them over and modify them to meet the needs of your own department. A user-friendly personal computer loaded with word processing software is also an essential tool. Word processors allow great flexibility when the time comes to make changes to and update SOPs. Obtaining someone else's SOPs on a disk in a word-processing language compatible with one your department has access to will save even more time.

Depending on the jurisdiction, the authority to issue policy and procedures may fall to the fire chief; the city or county executive officer; or the city council, board, or commission. In some jurisdictions, generic policies cover all municipal employees. Such general policies may be very broad and may not take into account the unique roles that public safety providers play.

In some jurisdictions, the environment may be so politicized that the chief lacks the authority to manage effectively. For example, in some volunteer departments, a fire chief may only discipline a member who disobeys the chief's orders if he has the vote of the full membership. If the member is popular, it may be difficult to obtain such a vote. When a fire chief doesn't have authority to issue and implement policies and procedures, it can be difficult to develop and manage an efficient, effective organization. Remember that, regardless of who has authority to develop policies and procedures, the fire chief will have to answer in the event of a lawsuit. No one should have absolute authority, but a chief must have the control necessary to be effective.

Involve everyone affected. Committees composed of the SOPs' end users—that is, employees or volunteer members—can be very useful in the development of SOPs. A group might be more successful in identifying and evaluating all of the issues involved. The downside, of course, is that the group may proceed in an entirely different direction from that which was originally intended. While this is not necessarily a bad thing, you can avoid possible complications by giving the group clear parameters from the outset.

A committee approach is especially helpful if the topic of an SOP is controversial or involves change. No one likes change. Therefore, it is desirable to get as many fingerprints on the proposed SOP as possible. This will help create a sense of ownership among committee members and may help in selling the final product to everyone affected.

The implementation phase. Alan Brunacini, chief of the Phoenix Fire Department and a noted writer and lecturer, suggests that department personnel develop procedures describing a standard course of action on the fireground before they develop SOPs. That is, first commit to writing how operations will be performed on the fireground. Second, lay out provisions for training and preparation prior to implementing a given SOP. Third, there should be an opportunity to test the application of an SOP. The application phase should be followed by a standard critique process. Finally, SOPs should be revised based on lessons learned during the application phase. Additional training should follow.[9]

Contents of your SOP manual. Develop the habit of brevity early in the process. The thicker your manual gets, the more likely it becomes that department members will ignore it. Keep the manual simple. If a procedure can be outlined in two pages, try to reduce it to one. Use simple, straightforward language. Keep ambiguities to a minimum. Where diagrams and illustrations are useful, include them. Using illustrations will help limit the length of individual documents. Of course, no matter how well an SOP is written, it will always be subject to interpretation. This is in part where grievance procedures, civil service boards, and arbitrators will come into play.

It is also important to remember that, while SOPs should instruct users on the tasks to be performed in a given situation, they should avoid telling *how* to perform that task unless absolutely necessary—for safety reasons, for instance, or because uniformity is essential in a given situation. For example, there are a number of different ways to load a preconnect hoseline, but it is in everyone's best interest if every member loads the hoseline the same way.

Format of your manual. When it comes to your format, almost limitless options are available. Borrowing another department's format can work well, although you should always make sure it hasn't been copyrighted.

A suggested outline for writing an SOP follows. It is based on the traditional Harvard outline format, which will probably look familiar to you. It is also easy to use. There is no right way to order or number subject headings—the complexity of the topic will determine a particular document's length. The only hard-and-fast rules to observe are to first indicate a given SOP's scope and purpose.

The following outline is based on the model set by the National Fire Protection Association (NFPA), which lists the scope and purpose at the beginning of each standard, guideline, and recommended practice it promulgates. This model is useful to follow because a typical policies and procedures manual will be based on NFPA standards. I like to conclude the document with the assignment of specific responsibilities. Remember that unless responsibilities are specifically spelled out, it can be difficult to hold members responsible for their actions. Assigning responsibilities protects both personnel and the fire chief.

The number of chapters or sections in your SOP manual and the subjects they address will depend on your department's size and mission and the scope of services provided. Local or state civil service regulations and bargaining agreements may make some of your manual's content more complex or may create additional rules and regulations. In such cases, it may be useful to create multiple volumes. For example, you may want to use separate volumes for the department's rules and regulations (along with civil service rules and your bargaining agreement) and the SOP manual.

Written EMS protocols are a must, given the litigation risks surrounding the practice of medicine. Emergency medical service protocols may either be included in your fire department manual or published separately. If your department provides advanced life support, those protocols should be promulgated by the department's operational medical director.

Arrangements of SOP manuals vary from department to department. At minimum, use headings that include the name of the department or agency, the title of the applicable section (e.g., Occupational Safety and Health), the page number, and the title of the standard or procedure.

Number rules, general orders, and policies and procedures for easy reference. Various numbering options are available to you. Some departments use the year and chronological sequence in which a procedure or order is issued—e.g., SOP 98-001. This scheme works well for numbering general orders. My personal preference for numbering SOPs, however, is to assign a unique three-digit number to each section or subject area, for example, as follows:

100 Rules and Regulations
200 General Administration
300 Hazardous Materials
400 Occupational Safety and Health
500 Maintenance
600 Emergency Operations
700 Emergency Medical Services
800 Communications
900 Fire Prevention

A slight modification of the above system will allow the promulgation of standards for up to 999 different categories. Of course, you'll never need that many sections.

Each category can then be divided into subheadings. For example, Section 500, Maintence, can be divided into the following subcategories:

500 General Maintenance
501 Apparatus and Motor Vehicles
502 Small Tools and Equipment
503 Fire Hose
504 Protective Clothing and Equipment
505 Ground Ladders
506 Generators and Electrical Equipment
507 Buildings and Grounds

Each three-digit subject identifier can be followed by a decimal point and two additional digits—e.g., SOP 503.02, Fire Hose Testing—for each procedure. Using this system allows you to assign 100 SOPs for each subheading. This should be more than adequate for even the most complex subjects.

When you begin writing your manual, you may want to alphabetize your subheadings within each category or subject area. As a practical matter, you will soon discover that your original list was not exhaustive, and subsequent procedures will fall in to the sequence as they develop. As an organization grows, so will its SOP manual. The early years of an organization will be marked by "must have" standards written in response to a crisis or in anticipation of some pressing need. As an organization matures, the procedures may begin to include the more subtle "nice to have" SOPs that close gaps and tie up loose ends. This is the reason a word processor is such a useful tool. It will allow you to periodically reorganize and renumber your list as needed.

Updating your SOPs. Standard operating procedures must be constantly reviewed and updated to remain effective. It is surprising how quickly five or ten years pass, rendering an SOP obsolete. If an SOP hasn't been revised in the past five years, it is probably no longer being used or has become unnecessary. If the former is true, you need to consider why the procedure has fallen into disuse and respond accordingly. If an SOP has become unnecessary, there no longer is any reason to include it in your manual. Issue dates and revision dates included in your headings will prove useful in reviewing your SOP manual with the intention of culling and updating it. Using month and year—e.g., 04/97—will suffice, although you might also consider using the date the SOP was issued.

SOPs Aren't Written in Stone

I am unaware of any policies or procedures that have been carved in stone. Developing a solid policies and procedures manual is a work in progress. The project is never complete. Once it is in usable form, distribute it in a looseleaf binder. To be effective and meaningful, policies and procedures must constantly be updated and corrected. New policies must be instituted and obsolete policies rescinded.

Preemption Issues

SOPs do not supersede policies and procedures of your municipality or other district. Rather, they amplify and augment them. It is useful to reference these documents in the applicable department policy or procedure. This will help limit the length of the SOP, since it will be unnecessary to repeat the policies verbatim.

Notes

For the sake of simplicity, the word *jurisdiction* is used throughout this book to refer to fire department jurisdiction. The words *county, city,* or *district* may be applicable to your department and should be used accordingly. References are also made throughout the book to various policies, programs, and governmental bodies, such as the Human Resources Handbook, Emergency Operations Plan, Emergency Management Program, fire council, and Department of Health. Your department does not exist in a vacuum, and a given SOP may be supplemented by provisions in a municipal code or a governmental department's policy. When promulgating SOPs, be sure your own procedures do not conflict with overriding applicable law or policy.

The minimum staffing levels and other numbers used herein are advisory. Do not construe them to exclude other possibilities. For instance, although a six-month probation period may be the norm, many departments have probationary periods of a year.

Masculine pronouns are used in this manual. However, there is a growing tendency to use gender-neutral language in official texts, and your department may opt to use "his or her" and "he or she" in its SOPs.

---

[1] Herbert J. Chruden and Arthur W. Sherman, *Personnel Management,* 3rd ed. (Dallas, TX: South-Western Publishing Company, 1968), 38.

[2] Ibid., 41.

[3] Henri Fayol, *General and Industrial Management* (Belmont, CA: David S. Lake Publishers, 1987), 57.

[4] Robert I. Mehr and Bob A. Hedges, *Risk Management in the Business Enterprise* (Homewood, IL: Richard D. Irwin, 1963), 16-26.

[5] Chruden and Sherman, 38.

[6] Alan V. Brunacini, *Fire Command* (Quincy, MA: National Fire Protection Association, 1985), 15-26.

[7] Ibid., 17.

[8] Alan V. Brunacini, "A Game Plan Reduces Legal Risk," *NFPA Journal* 86 (March/April 1992), 30.

[9] Alan V. Brunacini, "Critiques Can Make You Better," *Fire Journal* 85 (March/April 1991): 143-144.

Section 100

Rules and Regulations

Rules and Regulations, SOP 100.01

ADMINISTRATION

1. Each work area shall be assigned clipboards for posting general orders, special orders, and memos. The purpose of the clipboards shall be to facilitate communications and promulgate policy. A bulletin board shall also be provided for posting items of general correspondence. No item shall be posted without the approval of the fire chief or his designated representative.
2. A manual containing the department's official rules and regulations and standard operating procedures shall be provided in each work area.
3. Each supervisor shall be responsible for maintaining the clipboards, bulletin boards, and manuals described above.
4. Definitions:
   A. General correspondence: Letters of appreciation, meeting notices, and other items of an informational nature to be posted for review. Letters of appreciation and commendation shall be discarded after 30 days. Other items shall be discarded as appropriate.
   B. General orders: Consecutively numbered, written directives used to amend or clarify a policy or procedure and for information of a permanent nature. General orders shall be posted for review and kept in a permanent file. New general orders shall be read aloud to members at roll call by the officer in charge. Members shall also be required to initial new general orders to indicate that they have read the order.
   C. Memo: Consecutively numbered correspondence, generally of an informational nature. Memos may address administrative policies and alter or clarify routine practices but may not alter or amend an item addressed in the Rules and Regulations Manual or the SOP Manual. Memos shall be posted for review, if appropriate, and shall be maintained in a permanent file.
   D. Special order: A written, unnumbered directive that addresses a specific instance where a policy or procedure will be changed, altered, or amended for a specific period of time. Special orders will be posted during the specified time period and shall be discarded after they expire.
   E. Standard operating procedure: A written, numbered organizational directive that establishes a standard course of action.

Rules and Regulations, SOP 100.02

DEFINITIONS

Terms used in this manual shall have the following definitions:

1. Absent without leave (AWOL): Failure to report for duty without sufficient reason and without securing proper approval for leave in advance.
2. Acting: Serving temporarily in a position to which the member is not ordinarily assigned, usually in a position of higher rank.
3. Appeal: The right of a nonprobationary member to apply for review from any order, dismissal, or suspension by the fire chief.
4. Chain of command: The line of authority from the fire chief through a single subordinate at each level of command.
5. Compensatory time: The period of time during which a member is excused from active duty as compensation for hours worked in excess of the regular tour of duty. Also known as "exchange time."
6. County (or city or district): The physical area within the defined boundaries of the county (or city or district).
7. Days off: The time off granted to each member without loss of pay after the member completes his regular tour of duty.
8. Dismissal: The act of terminating the service of a member.
9. Eligibility list: A list of eligible candidates certified by the human resources department as having qualified to be considered for employment in an entry-level position.
10. Emergency callback: Callback to duty when emergency conditions require additional personnel to mitigate the emergency. Members shall be compensated for callback duty according to county policy.
11. EMS: Emergency medical service.
12. Fire code: Ordinance governing fire prevention as adopted by the county.
13. Funeral leave: The period of time during which a member is excused from active duty by reason of the death of an immediate family member.
14. Gender: Within this manual, the words "he" and "his" shall be construed to refer to both genders.
15. General bulletins: Consecutively numbered, written procedures used to clarify department policy or procedures or to disseminate information of a permanent nature. General bulletins shall be kept in a permanent file.
16. General order: Consecutively numbered, written directives used to change the department's rules, regulations, or standard operating procedures. General orders shall be kept in a permanent file.
17. Immediately: The term "immediately" shall be construed to mean "as soon as possible and practicable."
18. Incompetence: The inability to satisfactorily perform one's duties or responsibilities.
19. Injury on duty leave: The period of time during which a member is excused from duty by reason of being injured while on duty.
20. Inspection: The periodic exam of personnel, stations, or apparatus for appearance, readiness, fitness for duty, and attention to duty according to standards set out in the rules and regulations manual, standard operating procedures, and general orders.

21. Insubordination: The willful disobedience of any order, lawfully issued by a superior officer, or any disrespectful, mutinous, insolent, or abusive language toward a superior officer.
22. Length of service: The period of time starting from the date a member's employment begins until the present or until the date the member's employment ends.
23. May and should: The word "may" is permissive. "Should" is advisory. Where used, the word "should" implies that, while the procedure is not mandatory, it is in the best interest of everyone involved for the procedure to be followed.
24. Members: A collective term applied to all persons on the department's payroll.
25. Neglect of duty: Failure to give proper attention to the performance of one's duty.
26. Nonsworn employee: A civilian, nonuniformed employee.
27. Oath of office: The oath each member takes at the time he is commissioned into the department's service.
28. On duty: A member is on duty during the period of time when he is actively responsible for or engaged in the performance of his duties.
29. Off duty: A member is off duty on his days off and when on authorized leave and free of the responsibility of performing usual routine duties. Technically, a member is on duty at all times and may be subject to recall at any time.
30. Order: An instruction or directive, either written or oral, issued by a superior officer to a subordinate or group of subordinates in the course of duty.
31. Personnel: Fire department employees.
32. Plural words: Within this manual, singular words include the plural and plural words include the singular.
33. Probationary period: The initial six months (or other period of time) of new appointees' service, beginning with the date of employment.
34. Promotion: A change in a member's employment status to a position of greater responsibility or higher classification.
35. Promotion lists: A list of eligible candidates certified by the human resources department as having qualified for promotion.
36. Regular duty callback: Callback to duty to fill a vacancy on a shift when another member's absence leaves that shift below the minimum staffing level.
37. Rank: A grade of official standing. Each class of members of the department constitutes a rank.
38. Ranking officer: The officer having the highest rank in grade for the longest period of time, unless otherwise designated by competent authority.
39. Relieved of duty: An employment condition during which a member is not required or permitted to perform assigned duties but retains pay status. A member generally is relieved of duty when under investigation.
40. Reserve members: All persons in the department who provide complementary staffing without formal compensation.
41. Resignation: The act of voluntary termination of a member's service.
42. Retirement: Termination of a member's active service by reason of attainment of the statutory length of service and age requirements or because of an incapacitating disability.
43. Rules and regulations manual: A written collection of administrative policies, operational procedures, and rules and regulations authorized by an order of the fire chief.

44. Shall and will: The words "shall" and "will" as used herein indicate that the action referred to is mandatory.
45. Sick leave: The period of time during which a member is excused from active duty by reason of illness or injury that prevents the member from performing his duties.
46. Special bulletin: A written, unnumbered procedure covering a specific situation or event and that applies for a limited period of time.
47. Special duty: Any duty that requires a member to be excused from his regular duties.
48. Special order: A written, unnumbered directive covering a limited period of time during which the rules, regulations, or standard operating procedures will be changed. Special orders shall be kept in a permanent file.
49. Standby callback: A recall of off-duty members for standby duty in a station. Used during emergency conditions or during periods of peak activity.
50. Superior officer: Any member with supervisory responsibilities, either temporary or permanent, over members of a lower rank.
51. Suppression personnel: Members assigned to firefighting and emergency medical service response duties.
52. Suspension: An action taken whereby a member is denied the privilege of performing his duties as a consequence of dereliction of duty, breach of discipline, misconduct, or violation of regulations. Suspension is either the first step in the disciplinary process or the penalty assessed.
53. Sworn employee: A uniformed employee.
54. Tense: Words used in the present tense include the future tense.
55. Through official channels: Through the hands of the superior officer in the chain of command. Written and oral communications may be passed through interoffice mail or voice mail unless the urgent or sensitive nature of the matter requires personal face-to-face contact.
56. Tour of duty: The hours during which a member is on duty.
57. Vacation leave: The vacation time granted to all members of the department each year as established by the board of supervisors.
58. Workday: A tour of duty.

Rules and Regulations, SOP 100.03

ACTING IN A HIGHER CLASSIFICATION

1. Whenever a temporary vacancy exists in a classified position subject to the department's minimum staffing guidelines, the position shall be filled by another member of the department to maintain minimum staffing. If a member of the same rank is not available to fill a position, a member from the rank immediately below the vacant position may be appointed temporarily.
2. Temporary appointments to a higher classification shall be made from the eligibility list for that position.
3. If no eligibility list exists on the date of the appointment, the following guidelines shall be followed in making the appointment:
   A. The appointment shall be made from the rank immediately below the vacant position.

   B. No person shall be appointed as a driver unless he has successfully completed the driver's course and passed his most recent annual driving evaluation.
   C. The fire chief shall have the authority to make the temporary appointment after reviewing the qualifications of the members eligible for appointment.
4. A member performing the duties of a higher classification shall receive the base salary for the higher classification during the temporary assignment, provided that the member works in the higher classification for 30 or more consecutive days. See the applicable sections of the Human Resources Handbook.
5. When a temporary assignment ends, a member shall return to his previous position.

Rules and Regulations, SOP 100.04

EQUAL EMPLOYMENT OPPORTUNITY

1. Employment and promotions within the department shall be based on valid job-related needs and criteria.
2. Supervisory and employment-related decisions shall not be based on age, color, disability, ethnicity, national origin, political affiliation, race, religion, gender, or sexual orientation, unless such factors are directly related to bona fide position qualifications.
3. All members shall become familiar with and comply with the provisions of the jurisdiction's Equal Employment Opportunity Policies and Affirmative Action Plan. See the applicable sections of the Human Resources Handbook.

Rules and Regulations, SOP 100.05

DISCIPLINARY ACTIONS

1. Disciplinary action is a tool to allow supervisors to deal effectively with members whose performance or conduct is unacceptable.
2. Disciplinary actions are taken to promote the efficiency of department operations. In exercising discipline, the department will give due regard to each member's legal rights and will ensure that disciplinary actions are based on objective considerations without regard to age, color, disability, ethnicity, national origin, political affiliation, race, religion, gender, sexual orientation, or other nonmerit factors.
3. See the applicable section of the Human Resources Handbook.

Rules and Regulations, SOP 100.06

GRIEVANCE PROCEDURE

1. A grievance is a complaint or dispute by a member relating to employment, including the following:
   A. Disciplinary actions involving dismissal, demotion, or suspension, provided that dismissals are grievable whenever resulting from formal discipline or unsatisfactory job performance.
   B. The application of personnel policies, procedures, rules, and regulations.

C. Acts of retaliation resulting from the use of the grievance procedure, participation in the grievance of another member, compliance with any federal or state law, reporting any violation of such law to a governmental authority, or seeking any change in law before Congress or the state legislature.

D. Complaints of discrimination on the basis of age, color, disability, ethnicity, national origin, political affiliation, race, gender, or sexual orientation.

2. A member who believes he has a legitimate reason to file a complaint or grievance should consult the applicable section of the Human Resources Handbook.

Rules and Regulations, SOP 100.07

LEAVE AND VACATIONS

1. Members shall consult the applicable section of the Human Resources Handbook for specific details of the various types of leave approved by the county.
2. Any member determined to be absent without proper authorization shall be subject to disciplinary action.
3. No leave shall be taken until an Employee Leave Request Form is completed and approved by the member's supervisor. A copy of the form shall be provided to the member requesting leave.
4. Supervisors shall have the authority to approve or disapprove all forms of leave based on a member's leave balance and the department's minimum staffing needs. (Note: Local union contract also may apply in this situation.)
5. Minimum staffing for station duty shall be ________ members. If more than ________ members report for duty and sufficient members are available to staff all fire companies and ambulances as required, members may request time off for that shift and may be granted time off as staffing levels permit.
6. Minimum staffing levels for each division shall also be established. The staffing level shall be communicated to all affected members. Supervisors may not schedule leave if granting the leave will result in the division's working below minimum staffing level.
7. Paid sick leave is a benefit granted to members by the district and may be used whenever a member is unable to perform his duties due to illness or injury. Sick leave also may be used when a member has an appointment with a physician, is physically incapacitated, or is required to attend to an ill or injured spouse or minor child. (Some jurisdictions may not permit members to use sick leave to tend to a spouse or child.)
8. Whenever a member is unable to perform his job due to illness or injury, the member shall devote his full attention to recovery and shall not engage in any activity that might aggravate or prolong the illness or injury. Members shall remain at home for the duration of the illness or injury, except to the extent necessary to attend an appointment with a physician, obtain drugs or medication, or undergo therapy treatments. Other conditions may apply.
9. To receive paid sick leave, a member must notify his supervisor that he will be absent from work due to illness or injury prior to the beginning of his scheduled workday. Notification may be made by voice mail.
10. Supervisors shall monitor the use of sick leave by their subordinates to prevent misuse of this benefit.

Rules and Regulations, SOP 100.08

MEDIA RELATIONS

1. Statements to the media, news releases, and media campaigns must be approved by the fire chief or other authorized person prior to their release, except as provided below.
2. An incident commander is authorized to provide the media with general details concerning an incident.
3. Communications personnel are authorized to provide the media with a list of incidents. This information shall be limited to the dates, times, and locations of incidents.
4. Information pertaining to the cause and origin of an incident shall be released only by the fire marshal or the fire marshal's designated representative.
5. Information relating to personnel matters, department policy, department litigation, or other sensitive matters shall be released only by the fire chief.

Rules and Regulations, SOP 100.09

RESIGNATIONS

1. A member is requested to provide at least two weeks' notice of an intent to resign from the department, to allow ample time to process the notice. The fire chief may waive the notice requirement and allow the resignation to become effective immediately on receipt of a member's intention to resign.
2. Notice of resignation shall be in writing and shall be delivered to the member's immediate supervisor. The supervisor shall forward the notice to the applicable deputy chief for processing.
3. A resigning member should contact the appropriate county department for information related to benefit options.
4. A resigning member shall turn in all uniforms, pagers, keys, and other property issued by the department. A member may be assessed a replacement cost for any item that is not returned or is returned damaged.
5. See also the applicable section of the Human Resources Handbook.

Rules and Regulations, SOP 100.10

PROMOTIONS

Note: There are as many different promotional processes as there are fire departments. Processes may be established by state or local civil service laws or by union contract. The SOP offered here is one of many possibilities. Most important is that the procedure be in writing.

1. Promotional exams shall be competitive and open to all persons who have continuously held a position in the next lower classification or an equivalent position as specified by the fire chief for two years or more immediately prior to the date of the exam.
2. When an insufficient number of members in the next lower position qualify to take the exam, the fire chief may extend the exam to members with fewer than two years of service. If an insufficient number of personnel qualify even after this extension, the fire chief may extend the exam to members in the next lower

position, based on salary, to that for which the exam is to be held. The fire chief also may open the exam to external applicants.

3. At least 30 days prior to any promotional exam, notice shall be posted on a bulletin board located in the main lobby of headquarters and on the bulletin boards in the fire stations. The notice shall indicate the position for which the exam is to be held; the date, time, and place of the exam; the minimum qualifications established for the position by the fire chief; and the deadline for receiving applications.
4. All questions on the written portion of the promotional exam shall be taken from source materials listed in a notice posted at least 30 days prior to the exam. The source materials shall be made available to all members involved.
5. All applicants shall be given an identical written exam in the presence of each other. The preliminary results of the written portion shall be posted in the main lobby of headquarters within 72 hours of completion of the written exam.
6. The maximum grade possible on the written exam shall be 100 points. All persons scoring at least 70 points on the written exam shall be eligible to proceed to the performance assessment portion of the process.
7. Each applicant may, within five days of the posting of the written exam results, have one opportunity to examine his exam paper. If the member is dissatisfied with his grade, he may file a written appeal with the fire chief. The fire chief shall rule on the validity of any appeal within 10 days of receiving it.
8. The second phase of the promotional process is a performance test consisting of exercises designed to measure each member's ability to perform tasks relevant to the position. This part of the process may include tactical exercises and tests to measure the member's leadership capability, ability to perform administrative tasks, and communications skills. Each applicant will be graded by the assessment panel against a maximum possible score of 100 points.
9. An applicant's position on the final eligibility list shall be determined by averaging the scores of the written portion and the performance test. Unless exhausted, an eligibility list shall remain in effect for one year from the date that the list is certified by the appropriate oversight body. At the end of one year, the list shall expire.
10. When a vacancy occurs, the fire chief shall interview the three individuals on the eligibility list with the highest grades, and he shall be free to select the one he deems most qualified for the position. Anyone not selected shall retain his position on the list until the list expires or until the member is promoted.
11. All promoted persons who accept the promotion shall (or may) be required to pass an appropriate medical exam, a psychological exam, and a controlled substance screening prior to appointment.

Rules and Regulations, SOP 100.11

SENIORITY LIST

1. The deputy chiefs shall prepare seniority lists for their respective divisions. The seniority lists shall be kept up to date at all times.
2. The seniority list shall classify members according to rank. Members shall be listed by rank in descending order. Members shall be listed in their respective rank by time in grade. The member in a respective rank for the longest period of time shall appear at the top of the list for that rank.

3. If two or more members of a given rank have the same time in grade, their positions on the seniority list shall be based on their respective dates of employment. If their dates of employment are the same, their positions shall be determined by their ranking on the eligibility list for their current classification.
4. The seniority list shall include each member's date of employment and the date of the member's most recent promotion.

Rules and Regulations, SOP 100.12

SHIFT SWAPS

1. A shift swap is defined as that time when one member voluntarily works for another by their mutual agreement.
2. The district does not assume any responsibility for compensating a member who voluntarily agrees to work for another. Nor shall the extra hours worked by a member during a shift swap be used to determine payments for overtime, exchange time, or any other benefit.
3. Shift swaps shall be on a time-for-time basis only.
4. A member may swap shifts only with a member of equal rank and may not work in a higher or lower classification without approval of the fire chief.
5. When members agree to a shift swap, both members must complete and sign an Employee Leave Request Form. Before the members are relieved of their regular tours of duty, the members' supervisors also must approve the shift swap. Failure to obtain prior approval shall result in the members' being considered AWOL.
6. Once a shift swap has been approved, the member who agrees to work another's shift is responsible for working the agreed-on tour of duty. Any member who agrees to work another member's shift and who fails to report for duty at the appropriate time shall be subject to disciplinary action. Legitimate absences shall be charged to the appropriate category of leave (See SOP 100.07, Leave and Vacations). A member charged with leave for failure to work for another member shall be entitled to repayment of the leave time by the other member as if the shift swap had actually been worked.
7. Once an Employee Leave Request Form has been approved, the member originally scheduled to work is completely relieved of duty and shall receive full compensation as though he actually had worked his tour of duty.
8. In an emergency situation, a supervisor may give verbal permission for a shift swap. In such a situation, the member requesting the shift swap shall be fully liable for his scheduled tour of duty in the event the member agreeing to work fails to report for duty.
9. Where a supervisor gives verbal permission for an emergency shift swap, an Employee Leave Request Form shall be completed and filed as soon as possible.

Rules and Regulations, SOP 100.13

TEMPORARY APPOINTMENTS

1. The county administrator (or other appropriate official) has the authority to create temporary full-time positions for a period of time not to exceed 24 months or other appropriate period of time.

2. Appointments to temporary full-time positions shall be based on the eligibility list for that position, provided such a list exists at the time the appointment is made.
3. If no eligibility list exists from which to appoint a member to a temporary full-time position, the position shall be filled based on the following guidelines:
   A. Except for an entry-level position, the appointment shall be made from the rank or classification immediately below that of the vacant position.
   B. No member shall be appointed to a driver's position unless that member has successfully completed the driver's course and has passed his most recent annual driving evaluation.
   C. The fire chief shall have the authority to make the temporary appointment and shall do so after evaluating a member's qualifications for the position.
4. If no eligibility list exists for an entry-level position, the fire chief shall have the authority to fill the temporary position with part-time employees until an eligibility list can be developed and the selection process can be conducted. After an eligibility list is created, the temporary full-time positions shall be filled from that list.
5. Persons appointed to a temporary full-time position shall receive salary and benefits, except for retirement, life insurance, and education assistance, as though filling a regular full-time position.

Rules and Regulations, SOP 100.14

TRANSFERS

1. Every department member is subject to transfer as a result of a departmental reorganization or to meet operational needs.
2. A member mandatorily transferred shall be given as much prior notice as is practical to allow the member to make necessary personal arrangements.
3. When a member is transferred, all properly scheduled leave is transferred with him, regardless of whether it becomes necessary to pay overtime to maintain minimum staffing levels.
4. A member may request a transfer to another assignment provided he meets the position's minimum qualifications. To transfer, the member must submit a written request to the fire chief via the member's immediate supervisor.
5. Requests for transfers shall be accommodated whenever possible, but the fire chief has final authority to grant or deny a voluntary transfer request, based on the overall impact the transfer would have on the organization.
6. When a transfer request is denied, the fire chief shall provide the reasons for the denial to the member who made the request. The reasons shall be put in writing.

Rules and Regulations, SOP 100.15

WORKERS' COMPENSATION/INJURY LEAVE

1. A member injured while on duty shall complete an accident or injury report within 24 hours or as soon as possible after the accident/injury and shall forward it to his supervisor as soon as possible. The report shall set forth the details of the accident/injury, along with supporting documentation from the member's physician, if appropriate.

2. A member injured while on duty without fault or negligence on his part is entitled to a leave of absence at full pay for a maximum of 12 calendar months or until he is able to resume regular duties.
3. See the applicable section of the Human Resources Handbook for details.

Rules and Regulations, SOP 101.01

FIRE CHIEF—JOB DESCRIPTION

1. The fire chief is the executive head of the department and is directly responsible for the proper and efficient operation of the department.
2. The fire chief shall be appointed by the county administrator (or other appropriate official) and shall be answerable to the county administrator.
3. The fire chief shall be responsible for the following:
   - A. Supervising, regulating, and managing the department. The fire chief shall control all department activities.
   - B. Prescribing the specifications and manner of wear of uniforms and protective clothing and equipment.
   - C. Reorganizing any part of the department when in his judgment such reorganization would best serve the department.
   - D. Establishing rules, regulations, and procedures as necessary to ensure department efficiency and effectiveness.
   - E. Enforcing the department's rules, regulations, and procedures.
   - F. Reprimanding, preferring written charges, and suspending or dismissing members when conditions so warrant.
   - G. Preparing and submitting the department's annual budget recommendations to the county administrator.
   - H. Monitoring the expenditure of funds allocated to the department.

   Note: The following items may be included where applicable to local conditions.
4. The fire chief shall serve as an ex-officio member of the Fire and Rescue Commission.
5. The fire chief shall serve as the county's liaison to independent volunteer fire and rescue companies.
6. Nothing contained in these rules, regulations, and procedures shall be construed as limiting the power and authority granted to the fire chief by the county administrator and board of supervisors or by state and federal laws.

Rules and Regulations, SOP 101.02

DEPUTY CHIEF OF EMERGENCY SERVICES—JOB DESCRIPTION

1. The deputy chief of emergency services shall be appointed by the fire chief and shall be directly responsible to him for the proper and efficient operation of the department's emergency services and training divisions.
2. The deputy chief of emergency services shall be responsible for the following:
   - A. Supervising the deployment of personnel in fire and rescue stations assigned to perform fire suppression and emergency medical service response duties.

   B. Coordinating fire and emergency medical training for career and volunteer personnel.
   C. Serving as the department's liaison to the fire council.
   D. Preparing reports and maintaining records as required.
   E. Supervising the use, maintenance, and repair of the department's administration and training facilities.
   F. Acting in the capacity of fire chief in the fire chief's absence or when so instructed.
   G. Performing other such duties as may be required.

Rules and Regulations, SOP 101.03

DEPUTY CHIEF OF SUPPORT SERVICES—JOB DESCRIPTION

1. The deputy chief of support services shall be appointed by the fire chief and shall be directly responsible for the proper and efficient operation of the department's communications and emergency management divisions.
2. The deputy chief of support services shall be responsible for the following:
   A. Supervising the operation and maintenance of the enhanced 911, computer-aided dispatch (CAD), two-way radio, and paging systems.
   B. Supervising communications personnel.
   C. Supervising the development, exercise, and revision of the county's Emergency Operations Plan and Emergency Management Program.
   D. Supervising the operation and maintenance of the department's vehicle fleet and equipment.
   E. Preparing reports and maintaining records as required.
   F. Serving as the department's liaison to the communications division of the local law enforcement agency department.
   G. Acting in the capacity of fire chief when so instructed.
   H. Performing other such duties as may be required.

Rules and Regulations, SOP 101.04

DEPUTY CHIEF OF TECHNICAL SERVICES—JOB DESCRIPTION

1. The deputy chief of technical services shall be appointed by the fire chief and shall be directly responsible for the proper and efficient operation of the department's technical services division.
2. The deputy chief of technical services shall be responsible for the following:
   A. Serving as fire marshal for the jurisdiction.
   B. Supervising the activities of personnel assigned to the technical services division.
   C. Making departmental reports and referring official matters to the fire chief.
   D. Enforcing state laws and local ordinances pertaining to fire prevention and arson investigation.
   E. Directing and administering fire prevention inspection activities related to fire code enforcement.

F. Directing and monitoring the investigation of fires, explosions, fire deaths, false alarms, and other related malicious or criminal acts pertaining to fire.
G. Developing and conducting continuing comprehensive department and public education and information programs.
H. Evaluating, continually updating, and recommending improvements to the fire prevention program and the county's fire code.
I. Serving as a liaison to the county's building and development department and other agencies with fire and life safety responsibilities.
J. Acting in the capacity of fire chief when so instructed.
K. Meeting the minimum requirements for certification set forth by the department of fire programs (or other appropriate department or agency).
L. Performing other duties as assigned.

Rules and Regulations, SOP 101.05

CAPTAIN—JOB DESCRIPTION

1. Captain (or "company officer," "lieutenant," "sergeant," etc.) is the position immediately below that of deputy chief.
2. Captains shall be appointed through a competitive exam process that shall include both a written exam and a performance test. No member shall be appointed captain unless he meets, at minimum, the qualifications for certification at the Officer I level and has previously served at least two years as a driver.
3. Captains shall be responsible for the following:
   A. Commanding an engine or truck company and a fire or rescue station. This includes being responsible for the care, maintenance, and usage of all equipment and items pertaining thereto.
   B. Preparing reports and maintaining records as required.
   C. Conducting and supervising training for station personnel.
   D. Obeying, supporting, and enforcing the department's rules, regulations, policies, and procedures and requiring the same of his subordinates.
   E. Setting a good example for subordinates and requiring subordinates to meet all required standards of conduct and performance.
   F. Promptly reporting in writing any violation of the department's rules, regulations, policies, and procedures, and referring all official matters to the fire chief through the established chain of command.
   G. Supervising and assisting subordinates in the performance of maintenance, training, company inspections, public education programs, and emergency response activities.
   H. Performing other such duties as may be required.

Rules and Regulations, SOP 101.06

DRIVER—JOB DESCRIPTION

1. Driver is the position immediately below that of captain.
2. Drivers shall be appointed through a competitive exam process that shall include both a written exam and a performance test. No member shall be appointed driver unless he meets, at a minimum, the qualifications for driver set out in NFPA 1002, *Standard for Fire Department Vehicle Driver/Operator Professional Qualifications,* and has served a minimum of two years as a firefighter/EMT or firefighter/paramedic.
3. A driver shall be directly responsible to his captain for the following:
   A. Driving apparatus, operating pumps and aerial devices, performing inspections of apparatus, conducting fire prevention inspections of businesses, taking part in public education activities, and participating in firefighting and emergency medical duties.
   B. Possessing a thorough knowledge of how to operate equipment assigned to apparatus in accordance with current training policies.
   C. Possessing a thorough knowledge of the streets, hydrants, static water sources, and target hazards within his first-due area.
   D. Maintaining the apparatus and equipment to which he is assigned.
   E. Acting in the capacity of captain when properly instructed to do so.
   F. Performing other such duties as assigned.

Rules and Regulations, SOP 101.07

FIREFIGHTER/EMT (FIREFIGHTER/PARAMEDIC)—JOB DESCRIPTION

Note: Hiring practices vary by region and jurisdiction. Many urban departments hire inexperienced recruits and put them through an academy. The description here is written for a small- to medium-size department that provides EMS in addition to fire suppression.

1. Firefighter is the position directly below that of driver. The position of firefighter is the entry-level position in the emergency services division.
2. Firefighter/EMTs shall be appointed through a competitive exam process. No person shall be appointed firefighter unless eligible for certification as a Firefighter I and EMT. As a condition of employment, a firefighter is required to become a paramedic as soon as practical. Maintaining paramedic status is a condition of continued employment.
3. Firefighters are directly responsible to their captains (company officers) for the following:
   A. Participating in company inspections, public education activities, and emergency response operations.
   B. Being prepared to temporarily assume the duties and responsibilities of driver, if necessary.
   C. Possessing thorough knowledge of the operation of the apparatus and equipment to which they are assigned.

D. Possessing a thorough knowledge of the streets, hydrants, static water sources, and target hazards in their first-due area.
E. Maintaining and caring for all equipment assigned to them, their station, and their apparatus.
F. Acting in the capacity of driver when properly instructed to do so.
G. Performing other such duties as assigned.

Rules and Regulations, SOP 101.08

ASSISTANT FIRE MARSHAL—JOB DESCRIPTION

1. Assistant fire marshal is the position immediately below that of deputy chief of technical services. The assistant fire marshal shall hold the rank of captain.
2. The assistant fire marshal shall be appointed by the deputy chief of technical services. The appointment shall be based on the results of a competitive process involving both a written exam and a performance test. No person shall be appointed assistant fire marshal who has not previously served for at least two years as an investigator or inspector.
3. The assistant fire marshal shall be directly responsible to the deputy chief of technical services for the following:
   A. Supervising and directing the activities of employees assigned to his command.
   B. Setting a good example for employees and requiring that subordinates meet all required standards of conduct and performance.
   C. Conducting fire prevention inspections of businesses, schools, day care facilities, hospitals, nursing homes, and other occupancies as assigned.
   D. Assisting in the investigation of fires and explosions.
   E. Investigating complaints and conducting internal affairs and background investigations as assigned by the fire chief.
   F. Presenting fire prevention programs and demonstrations.
   G. Conducting public education programs as requested.
   H. Completing reports, taking statements, performing filing, and maintaining case folders and records of inspections and investigations.
   I. Possessing the minimum certifications required by the department and other designated agencies.
   J. Performing other such duties as assigned.

Rules and Regulations, SOP 101.09

INVESTIGATOR—JOB DESCRIPTION

1. Investigator is the position immediately below that of assistant fire marshal and equivalent to that of driver in the emergency services division.
2. Investigators shall be appointed by the deputy chief of technical services based on a competitive process that involves both a written exam and a performance test.
3. Investigators are responsible to the assistant fire marshal for the following:
   A. Investigating fires and explosions, determining cause and origin, preparing reports detailing cause and estimated losses, and obtaining and preserving evidence.

B. Apprehending and prosecuting arson offenders and filing cases with the appropriate government entity—e.g., the state attorney's office.
C. Conducting fire prevention inspections and enforcing the provisions of the county's fire code.
D. Preparing and maintaining reports and records as required.
E. Providing consultation for emergency drills.
F. Developing and presenting fire prevention programs.
G. Counseling juveniles on the dangers of false alarms and other fire safety-related actions.
H. Performing courtesy home fire and safety inspections on request.
I. Meeting the department's minimum certification requirements and the requirements of other appropriate agencies.
J. Maintaining the certifications and/or licenses required to lawfully carry a firearm (See SOP 900.01, Weapons).
K. Cleaning and maintaining the issued firearm in accordance with SOP 900.01, Weapons.
L. Performing other such duties as assigned.

Rules and Regulations, SOP 101.10

INSPECTOR—JOB DESCRIPTION

1. Inspectors hold the rank immediately below that of investigator, equivalent to that of firefighter in the emergency services division.
2. Inspectors shall be appointed by the deputy chief of technical services based on a competitive process that includes both a written exam and a performance test.
3. Inspectors shall be responsible to the assistant fire marshal for the following:
   A. Conducting fire prevention inspections in businesses, schools, day care facilities, hospitals, nursing homes, and other occupancies as assigned.
   B. Assisting in the investigation of major incidents.
   C. Presenting fire prevention programs and demonstrations.
   D. Conducting public education programs as requested.
   E. Completing reports, taking statements, filing appropriate forms, and maintaining case folders and records of inspections and fires.
   F. Meeting the department's minimum certification requirements and the requirements of other appropriate agencies.
   G. Performing other such duties as assigned.

Rules and Regulations, SOP 101.11

TRAINING CAPTAIN—JOB DESCRIPTION

1. Training captain is the position immediately below that of deputy chief of emergency services.
2. The training captain shall be appointed by the deputy chief of emergency services based on a competitive process involving both a written exam and a performance test. No person shall be eligible to be appointed training captain unless he has previously served a minimum of two years as a training officer or driver.

3. The training captain shall be directly responsible to the deputy chief for the following:
   A. Developing, conducting, coordinating, and managing fire and emergency medical training programs.
   B. Supervising and working with the training officers in the development and delivery of training programs.
   C. Setting a good example for his subordinates and ensuring employees meet all required standards of conduct and performance.
   D. Scheduling training programs and managing the use and maintenance of the department's training apparatus, vehicles, and fixed facilities.
   E. Serving as a liaison to the various volunteer or other appropriate committees involved in training activities.
   F. Maintaining a permanent training file for each member in the system and preparing required reports.
   G. Possessing the minimum certifications required by the department and other appropriate agencies.
   H. Attending training programs and seminars and reading periodicals and journals to stay knowledgeable of current industry trends and practices.
   I. Performing other duties as assigned.

Rules and Regulations, SOP 101.12

TRAINING OFFICER—JOB DESCRIPTION

1. Training officer is the position immediately below that of training captain and equivalent to that of driver.
2. The training officer shall be appointed by the deputy chief of emergency services based on a competitive process involving both a written exam and a performance test. No person shall be eligible for appointment as training officer unless he has served a minimum of two years as a firefighter/EMT.
3. The training officer shall be directly responsible to the training captain for the following:
   A. Developing, conducting, coordinating, and supervising fire and emergency medical training classes, courses, and seminars.
   B. Setting a good example for his students and requiring that they meet all required standards of conduct and performance.
   C. Maintaining records and preparing reports as required.
   D. Attending training courses, seminars, and conferences and reading trade journals to keep up with industry trends and changes.
   E. Maintaining skill level through periodic assignment to response duties.
   F. Supervising and maintaining assigned work areas.
   G. Possessing the minimum certifications required by the department and other appropriate agencies.
   H. Performing other such duties as assigned.

Rules and Regulations, SOP 101.13

COMMUNICATIONS SUPERVISOR—JOB DESCRIPTION

1. Communications supervisor is the position immediately below that of deputy chief of support services and equal to that of captain in the emergency services division.
2. The communications supervisor shall be appointed by the deputy chief of support services based on a competitive process involving both a written exam and a performance test. No person shall be appointed communications supervisor unless he has served a minimum of two years as a communications shift supervisor or an equivalent position within the department.
3. The communications supervisor shall be directly responsible to the deputy chief of support services for the following:
   A. Supervising the communications division in the receipt and dispatch of all emergency and nonemergency telephone calls received for fire and EMS.
   B. Maintaining records, submitting reports, and referring all official matters to the deputy chief of support services.
   C. Training subordinates.
   D. Obeying, supporting, and enforcing the department's rules, regulations, policies, and procedures and requiring the same of his employees.
   E. Setting a good example for subordinates and requiring that employees meet all required standards of conduct and performance.
   F. Immediately reporting to his supervisor any deficiency in equipment, personnel, or records that makes adequate performance of duties impossible.
   G. Supervising the use and maintenance of the computer-aided dispatch (CAD) system.
   H. Creating from the CAD records and reports as required.
   I. Performing other such duties as assigned.

Rules and Regulations, SOP 101.14

COMMUNICATIONS SHIFT SUPERVISOR—JOB DESCRIPTION

1. Communications shift supervisor is the position immediately below that of communications supervisor and equal to that of driver in the emergency services division.
2. Communications shift supervisors shall be appointed by the deputy chief of support services based on a competitive process involving both a written exam and a performance test. No person shall be appointed communications shift supervisor unless he has previously served a minimum of two years as a dispatcher or an equivalent position within the department.
3. Communications shift supervisors shall be directly responsible to the communications supervisor for the following:
   A. Supervising and assisting their shifts in receiving and dispatching all emergency and nonemergency telephone calls received for fire and EMS.
   B. Maintaining records, submitting reports, and referring all official matters to the communications supervisor.

C. Training their subordinates.
D. Obeying, supporting, and enforcing the department's rules, regulations, policies, and procedures and requiring the same of their employees.
E. Setting a good example for their subordinates and requiring that employees meet all required standards of conduct and performance.
F. Immediately reporting to their supervisor any deficiency in equipment, personnel, or records that makes adequate performance of duties impossible.
G. Performing other duties as assigned.

Rules and Regulations, SOP 101.15

DISPATCHER—JOB DESCRIPTION

1. Dispatcher is the position immediately below that of communications shift supervisor and is the entry-level position for the support services division.
2. Dispatchers shall be appointed by the deputy chief of support services based on a competitive process involving both a written exam and a performance test.
3. Dispatchers shall be directly responsible to the communications shift supervisor for the following:
   A. Receiving and recording all emergency and nonemergency telephone calls received for fire and EMS.
   B. Dispatching appropriate apparatus to emergency and nonemergency incidents.
   C. Initiating written chronological records of fire and EMS incidents.
   D. Ensuring that all emergency telephone and radio traffic is recorded on electronic recording equipment.
   E. Maintaining a current, up-to-date street index and up-to-date maps.
   F. Immediately reporting to their supervisor any deficiency in equipment, personnel, or records that makes adequate performance of duties impossible.
   G. Performing other duties as assigned.

Rules and Regulations, SOP 101.16

EMERGENCY MANAGEMENT COORDINATOR—JOB DESCRIPTION

1. Emergency management coordinator is the position immediately below that of deputy chief of support services and is equivalent to that of captain in the emergency services division.
2. The emergency management coordinator shall be appointed by the deputy chief of support services based on a competitive process involving both a written exam and a performance test. No person shall be appointed emergency management coordinator unless he has previously served a minimum of two years as a driver or an equivalent position within the department.
3. The emergency management coordinator shall be directly responsible to the deputy chief of support services for the following:
   A. Preparing, maintaining, and updating the district's Emergency Operations Plan.

B. Preparing, conducting, and coordinating exercises to test the county's readiness for man-made and natural disasters.
C. Serving as the department's liaison to the various local, state, and federal agencies with an interest in emergency management.
D. Maintaining records and preparing required reports.
E. Developing, coordinating, and participating in public education programs to increase public awareness of emergency management issues.
F. Possessing the minimum certifications required by the department and the state's department of emergency services.
G. Performing other such duties as assigned.

Rules and Regulations, SOP 101.17

ADMINISTRATIVE MANAGER—JOB DESCRIPTION

1. The administrative manager shall be appointed by the fire chief and shall serve as a member of the chief's staff.
2. The administrative manager shall be directly responsible to the fire chief for the following:
   A. Assisting in the preparation of the department's budget, managing procurement, and monitoring expenditures on a daily basis.
   B. Coordinating the submission of items to the appropriate governing body—e.g., the board of supervisors.
   C. Serving as a liaison between the department and the department of information technology (computer services or other local agency).
   D. Administering contracts.
   E. Maintaining records, managing the department's information management system, and preparing reports as required.
   F. Serving as the department's human resources coordinator.
   G. Serving as the department's public information officer.
   H. Performing other duties as required.

Rules and Regulations, SOP 101.18

SECRETARY—JOB DESCRIPTION

1. Secretaries shall be appointed by the fire chief.
2. Each secretary shall be assigned to a specific division and/or supervisor.
3. Secretaries shall be directly responsible to their assigned supervisor for the following:
   A. Maintaining records, reports, and files as required.
   B. Acting as receptionists and providing information to the public and staff as requested.
   C. Being proficient in the use of all office equipment as required, such as typewriters, fax machines, computers, copiers, calculators, and so forth.
   D. Possessing a thorough knowledge of business English; shorthand and dictation; and local ordinances, rules, regulations, policies, practices, procedures, and the services provided by county departments.

E. Being proficient in spelling, basic mathematics, and proper telephone techniques.
F. Maintaining confidentiality and trust placed in them by their respective supervisors.
G. Performing other such duties as required.

Rules and Regulations, SOP 101.19

VOLUNTEER COORDINATOR—JOB DESCRIPTION

1. The volunteer coordinator shall be appointed by the fire chief and shall serve as a member of the chief's staff.
2. The volunteer coordinator shall be directly responsible to the fire chief for the following:
   A. Serving as the department's liaison to the appropriate organizations and agencies.
   B. Coordinating and assisting with the administration of the volunteer benefits program.
   C. Editing and publishing the department's newsletter.
   D. Advising the fire chief on matters pertaining to the volunteer system.
   E. Actively participating in the department's public education program.
   F. Coordinating the development and management of a systemwide safety officers' association.
   G. Performing other such duties as required.

Rules and Regulations, SOP 102.01

CODE OF CONDUCT

1. As a basic condition of membership, all members have an obligation to conduct their official duties in a manner that serves the public interest, upholds the public trust, and protects the department's resources. To this end, all members have the responsibility to:
   A. perform their duties to the very best of their abilities and in a manner that is efficient, is cost-effective, and meets the needs of the public;
   B. demonstrate integrity, honesty, and ethical behavior in the conduct of all department business;
   C. ensure that personal interests do not come in conflict with official duties and avoid both actual conflicts of interest and the appearance of conflicts of interest when dealing with vendors, customers, and other individuals doing business or seeking to do business with the department;
   D. ensure that all department resources, including funds, equipment, vehicles, and other property, are used in strict compliance with department policies and solely for the benefit of the department;
   E. conduct all dealings with the public, county employees, and other organizations in a manner that presents a courteous, professional, and service-oriented image of the department;
   F. treat the public and other employees fairly and equitably, without regard to age, color, disability, ethnicity, national origin, political affiliation, race,

religion, gender, sexual orientation, or any other factor unrelated to the department's business;

G. avoid any behavior that could fall under the definition of misconduct in the disciplinary section of the Human Resources Handbook; and

H. report for duty at the appointed time and place fully equipped, fit, and able to perform assignments.

2. Officers and supervisors shall set an example for other members and have a responsibility to ensure that their activities and decisions pertaining to community services, personnel actions, and the management of public funds are consistent with the department's policies and practices.

Rules and Regulations, SOP 102.02

OBEDIENCE TO ORDERS

1. Members shall read and become familiar with the department's rules, regulations, policies, and procedures. No plea of ignorance of the rules and regulations will be accepted as an excuse for any violation.
2. Members shall promptly and willingly respond to the lawful orders of superior officers or acting officers. Refusal to obey a lawful order shall constitute insubordination. Obvious disrespect for or disruption of a supervisor's order likewise shall be deemed insubordination.
3. Members shall abide by federal and state law, local ordinances and rules, and the department's general orders and rules of conduct. Members shall not be required to obey orders that are illegal or in conflict with the department's rules and regulations.
4. Members shall not publicly criticize or comment derogatorily to anyone about instructions or orders received from a superior officer.
5. Supervisors and acting supervisors shall refrain from exceeding their authority in giving orders. The wrongful or injurious exercise of authority is prohibited.
6. Every officer, on and off duty, will be held responsible for enforcing the department's rules. If a violation comes to an officer's attention, that officer shall immediately notify the member of the violation and take corrective action. Should an officer fail to report a violation of an order or the department's rules, that officer shall be equally responsible for the violation.
7. Should a member receive an order that conflicts with a previous order, the member shall notify the officer who issued the conflicting order and shall be governed by the officer's subsequent instructions.
8. Any member who is given an order he believes to be unjust, improper, or contrary to a general order or rule of the department or a federal, state, or county policy should respectfully decline to obey the order and shall state the reason for doing so. The member shall request that the supervisor of the person issuing the order be contacted for instructions if the person issuing the order does not rescind or alter the original order.
9. A member may appeal for relief from orders or instructions that the member believes to be illegal, unjust, or improper. See the appropriate section of the Human Resources Handbook pertaining to the county's policy on this matter.

Rules and Regulations, SOP 102.03

PROFESSIONAL RELATIONS

1. Department members shall exhibit courtesy and respect to all officers and acting officers. While on duty, all officers shall be referred to by their appropriate rank.
2. Supervisors shall exhibit courtesy and respect to their subordinates and shall treat all members in a fair and impartial manner.
3. Members shall treat one another with due courtesy and shall not engage in horseplay or disrespectful conduct while on duty.
4. Members are required to speak the truth at all times, whether or not under oath, in giving testimony, in connection with official orders, and in connection with official duties.
5. Members shall not make false reports concerning any department business or the personal character or conduct of any member.
6. Members shall exhibit courtesy and respect to members of the public and other county employees.
7. Members are required to give their name and rank whenever requested by a member of the public.
8. Should a member have a complaint against a member of the public, he shall forward the complaint in writing to the fire chief.

Rules and Regulations, SOP 102.04

PERSONAL APPEARANCE

1. This rule applies to all members while on duty or officially representing the department at a public meeting, training session, seminar, conference, or other similar event.
2. Members shall maintain proper personal hygiene while on duty.
3. Uniforms and shoes shall be neat and clean and shall conform to the requirements set forth in SOP 102.08, Uniforms.
4. When not in uniform, members who are on duty or who are representing the department shall dress in a professional manner that is appropriate for the occasion. At no time while in uniform shall a member wear jewelry, pins, ribbons, buttons, or an article of clothing that constitutes an advertisement; a religious, political, or social viewpoint; or a message that is offensive to anyone on the basis of age, color, disability, ethnicity, national origin, race, religion, political affiliation, gender, or sexual orientation.
5. Hair shall be kept clean and well-groomed, shall not constitute a safety hazard, and at no time shall interfere with the use of protective clothing or equipment.
6. Beards and goatees are prohibited. Neatly trimmed sideburns and mustaches are permitted provided they do not interfere with the use of protective clothing or equipment. Sideburns shall not extend below the base of the ear and mustaches shall not extend below the bottom lip.
7. Members of the emergency services division shall limit their use of jewelry to a wrist or pocket watch, a wedding ring, and one school or university ring. These items shall not interfere with the proper use of protective clothing or equipment. Earrings, ear studs, bracelets, neck chains, and so forth are prohibited. (Local policy may vary.)

8. Members who are not involved in emergency response operations shall limit their use of jewelry. Items that create excessive noise, interfere with job performance, or constitute a safety hazard shall not be worn.
9. Male department members are prohibited from wearing earrings and ear studs. Female members, other than those in the emergency services division, shall limit their use of earrings and ear studs to a single pair. Nose jewelry or other items that draw attention to the wearer are also prohibited. (Local policy may vary.)

Rules and Regulations, SOP 102.05

PHYSICAL AND MENTAL FITNESS

1. All members are subject to the preemployment, retention, release-to-work, and reevaluation physical examinations required by the appropriate section of the Human Resources Handbook.
2. Each member shall remain mentally and physically fit and shall be able to perform his job duties.
3. When appropriate, members are encouraged to take advantage of the services provided by the county's Employee Assistance Program.

Rules and Regulations, SOP 102.06

RECALL TO DUTY

1. All members shall remain at work until properly relieved of duty.
2. To maintain essential services, the fire chief or his designee shall have the authority to order members of the department to return to duty at times other than their normal work period.
3. Members recalled to duty shall be compensated in accordance with the appropriate section of the Human Resources Handbook. Compensation will begin from the time the employee reports for duty.
4. Members recalled to duty shall report within a reasonable period of time after being notified to report to their designated duty site.
5. Members shall respond to an emergency recall unless incapacitated. Any member who refuses to respond shall be subject to disciplinary action for insubordination.

Rules and Regulations, SOP 102.07

STATION DUTY

1. Captains (company officers) are responsible for ensuring that the tasks assigned according to the daily work schedule are completed in a timely manner and that their apparatus and station are clean and the company capable of responding to emergencies.
2. Crews are expected to turn out within 60 seconds of receipt of an alarm.
3. Captains may waive the tasks assigned according to the daily work schedule when special circumstances warrant. Any activity not completed in such a situation shall be rescheduled for the next tour of duty and shall be completed as soon

as practical. Special circumstances may require that a crew continue to work after their normal relief time to complete an assignment.

4. Captains shall complete a Daily Activity Report for each tour of duty. The completed report shall be forwarded to the fire chief.
5. Drivers shall complete a Driver's Daily Apparatus Checklist for their assigned vehicles for each tour of duty.
6. Prior to being relieved of duty, members shall report all pertinent information to the crew relieving them.
7. Protective clothing shall be stored properly at the end of each tour of duty and shall not be left on the apparatus.
8. Members shall not watch television or engage in recreational activity except during meal breaks limited to one hour and after. Sleeping is not permitted by members on duty until after __________ hours. (This requirement may be waived during a disaster or a period of exceptional incident volume.)
9. Captains are responsible for station operations and shall take appropriate actions to ensure that fuel, utilities, and station supplies are used conservatively.
10. Whenever the daily schedule permits, members are encouraged to devote their free time to physical fitness activities and personal study. However, members may not engage in any activity that interferes with their ability to respond promptly to an incident.

Rules and Regulations, SOP 102.08

UNIFORMS

1. General requirements.
   - A. All members shall be issued uniforms and shall wear the appropriate uniform while on duty. This rule does not apply to the fire chief, the deputy director of technical services, fire investigators, and support staff.
   - B. Uniforms shall not be worn off duty except as provided in this section.
   - C. Uniformed members of the department shall report for duty in the prescribed uniform.
   - D. Only uniforms issued and approved by the department shall be worn while on duty or at other times identified herein as appropriate.
   - E. No part of a uniform shall be worn with nonuniform garments, nor shall nonuniform items be worn with the uniform.
   - F. Uniforms shall be kept clean, neat, and in a proper state of repair. Faded, frayed, and worn-out items shall not be worn but shall be turned in to the uniform officer for proper disposal.
   - G. Jacket and shirt pocket flaps shall be buttoned at all times. Pockets shall be free of objects that create bulges or that otherwise detract from a professional appearance.
   - H. A member in uniform shall at all times wear a belt. Only department-issued belts may be worn with the uniform.
   - I. Shoes and boots shall be kept clean and polished at all times. Only department-issued shoes and boots shall be worn by members with emergency response duties. Uniform shoes worn by members without emergency response duties shall be black leather lace-type or Wellington style (smooth,

round-toed, pull-on) boots. Tennis shoes, loafers, and other types of casual footwear may not be worn.

J. Sleeves shall be buttoned at the wrist and shall not be rolled up.

2. Uniform types.

A. *Class A, full-dress uniform.* Class A uniforms shall be worn at funerals, award ceremonies, parades, and other events as ordered by the fire chief. The uniform shall consist of:

1. navy blue, double-breasted coat and matching trousers;
2. long-sleeved shirt, white for officers and light blue for all others;
3. black tie and black belt;
4. black lace-up shoes or Wellington-style boots with black socks;
5. dress hat, badge, and rank insignias;
6. medals or award ribbons centered above the nameplate;
7. gold nameplate for officers and silver nameplate for other members, centered on the right side of the uniform with the bottom edge touching the top edge of the breast pocket; and
8. plain white undershirt/undergarments.

B. *Class B, station uniforms.* Class B uniforms shall be the standard attire unless otherwise directed. The uniform shall consist of:

1. for emergency response personnel, navy blue shirt and trousers that comply with NFPA 1975, *Station / Work Uniforms for Fire Fighters;*
2. for nonemergency response personnel, navy blue shirt and trousers (the garments need not meet NFPA 1975 requirements);
3. short-sleeved shirts from 1 April to 31 October, long-sleeved shirts from 1 November to 31 March;
4. black belt;
5. black or white socks (local preference may vary);
6. department-issued navy blue T-shirt;
7. baseball cap; and
8. department-issued shoes or boots that comply with ANSI Z41, *Standard for Safety-Toe Footwear,* or NFPA 1974, *Protective Footwear for Structural Fire Fighting.*
9. Rank insignias, badges, nameplates, medals, and award ribbons shall not be worn with the class B uniform. However, members of the technical services division shall wear an embroidered badge. The name and rank of all members shall be embroidered on their uniform shirts.
10. The department's patch shall be worn on the left shoulder of the class B uniform. EMT and paramedic patches shall be worn on the right shoulder of the class B uniform. Patches shall be centered on the uniform sleeve. No other patches may be worn.

C. *Class C, work uniform.* The class C work uniform may be worn while performing maintenance and training activities. It shall consist of a department-issued T-shirt or sweatshirt in lieu of the class B shirt. In some instances, coveralls will be issued to members as a class C uniform.

3. Rank insignias

A. Gold rank insignias shall be worn on the class A uniform by the fire chief, deputy fire chiefs, division commanders, captains, training captains, assistant fire marshals, communications supervisor, and emergency management coordinator.

B. The officer's rank insignia shall be sewn on the collar of the class B uniform.

C. All other members shall wear silver department rank insignias on their class A uniforms and no rank insignia on their class B uniforms.

D. The insignia shall be worn parallel to the front edge of the collar, centered one-half inch from the top and bottom.

4. Physical fitness clothing

A. Members shall wear department-issued clothing while involved in personal physical fitness training. The clothing shall include T-shirts, shorts, and sweatpants.

B. Members shall not wear unauthorized clothing during physical fitness training.

C. Members shall don protective clothing, coveralls, or uniforms prior to responding to a call during a workout activity.

5. Seasonal clothing

A. The department shall issue jackets, coats, and rainwear as appropriate.

B. Navy blue or white thermal underwear and navy blue, long-sleeved T-shirts may be worn with the uniform as appropriate.

6. Nonuniformed personnel

A. The fire chief, deputy chief of technical services, fire investigators, and support staff may wear civilian clothing.

B. All clothing worn by nonuniformed members shall be neat, clean, and appropriate for the occasion.

C. Nothing may be worn that could constitute a safety hazard or be offensive to another person on the basis of age, color, disability, ethnicity, national origin, political or social affiliation, race, religion, gender, or sexual orientation.

Rules and Regulations, SOP 102.09

VEHICLES AND EQUIPMENT

1. General requirements

A. Vehicles and equipment shall be maintained in a constant state of readiness and availability for complete and immediate use.

B. Members shall be responsible for the use and care of vehicles and equipment assigned to them or entrusted to their care.

C. Members shall immediately report any loss, damage, or malfunction of apparatus or equipment to the member's supervisor. Damage, destruction, or loss due to the member's negligence may result in his being required to make restitution. Disciplinary action also may be taken as appropriate. (Consult with your department's attorney prior to requiring restitution or taking disciplinary action.)

D. Members shall return vehicles and equipment issued to them or entrusted to their care immediately on separation from service.

E. A member required to drive a vehicle owned or operated by the department shall possess an appropriate and valid driver's license.

F. Members shall drive in a safe and prudent manner and shall obey all applicable federal, state, and local traffic regulations when driving or operating a vehicle owned or operated by the department.

G. Members shall properly wear safety restraint devices whenever driving or riding in a vehicle owned or operated by the department.

H. Members shall not use tobacco products while driving or riding in a vehicle owned or operated by the department.

2. Use of department vehicles

   A. Vehicles owned or operated by the department shall be used for county business only. County business means any authorized work or activity performed by a member on behalf of the county.

   B. An officer may authorize a brief stop at a convenience store or other similar establishment for a break while his company is within its district performing an authorized activity. The company must maintain radio contact and remain available for calls.

   C. Department vehicles may be used to procure meals or groceries for station meals. When obtaining groceries, a company must do the following:

      1. Maintain radio contact and remain available for calls.
      2. Send only one crew member (with a portable radio) into the store to procure the supplies.
      3. Make only one trip per shift.

3. Taking vehicles home

   A. The following members are authorized to take a vehicle home: the fire chief, deputy chief of emergency services, deputy chief of technical services, deputy chief of support services, fire investigators subject to callback, emergency management coordinator, and fire and EMS training captains.

   B. When circumstances warrant, the fire chief may authorize other members to take a vehicle home on a case-by-case basis.

4. Motor pool

   A. The department shall maintain a motor pool of vehicles for use by members while on duty.

   B. The vehicles in the motor pool shall be staged at the department's headquarters facility, and the keys shall be kept by the deputy chief of support services.

   C. If a motor pool vehicle is unavailable, the director may authorize a member to use his personal vehicle. Members shall be reimbursed according to county policies and procedures.

5. Injuries and property damage

   Any accident or collision involving damage to any vehicle or property or injury to any person shall be reported immediately to the appropriate law enforcement agency and to the member's supervisor.

Rules and Regulations, SOP 102.10

VISITORS AT STATIONS

1. Members are permitted to have visitors at their place of work.
2. Visitors are not permitted to enter a station dormitory or locker room unless it is necessary to access a restroom or other station facility.

3. Visitors are not allowed to enter a workshop or apparatus bay unless properly escorted by a department member.
4. When escorting visitors in an apparatus bay, members shall not allow children to play on, around, or with emergency apparatus or equipment unless carefully and closely supervised.
5. Visitors shall not be allowed to disrupt the daily work schedule. Their visits shall be limited to a maximum of 30 minutes.
6. Amorous activity with a visitor is not permitted while on duty.
7. Visitors are expected to abide by department rules and regulations while at fire stations or other work sites.
8. Minors shall at all times remain under the supervision and control of an adult.

Rules and Regulations, SOP 103.01

CONTROLLED SUBSTANCES

1. The use of alcoholic beverages, debilitating drugs, or any substances that impair physical or mental capabilities while on duty is strictly prohibited.
2. Off-duty consumption of alcohol that reflects negatively on the department or that impairs a member's ability to perform his job is prohibited.
3. Members shall be familiar with and strictly comply with the drug- and alcohol-free workplace provisions of the Human Resources Handbook.

Rules and Regulations, SOP 103.02

INAPPROPRIATE BEHAVIOR

The following activities are prohibited by members on duty.

1. Unlawful behavior, gambling, noisy or quarrelsome conduct, and lewd or indecent activity.
2. Possession of a firearm or other deadly weapon unless the member is authorized by the fire chief to carry such a weapon.
3. Threats or acts of physical violence against members of the public, coworkers, or other department members or county employees.
4. Sexual activity to include the possession or use of printed or audiovisual material that is sexually offensive.
5. Abusive behavior, hazing, or harassment of coworkers or members of the public. Horseplay, practical jokes, and other disruptive behavior is also prohibited.
6. Use of department supplies, tools, and materials to clean or repair personal vehicles or property.
7. Alteration or modification of vehicles, apparatus, buildings, computers, or items of equipment owned or operated by the department without the fire chief's authorization.
8. Acceptance or solicitation of gifts, rewards, or fees for services incidental to the performance of one's duty. In addition, no member shall be required to make a donation to any person or organization as a condition of employment.
9. Campaigning for or against any elected official.
10. Publicly criticizing the official actions or orders of a superior officer. Nor may a member publicly speak disrespectfully of the department or its members.

11. Recommending or endorsing specific products, trade names, or businesses.
12. Conducting personal business or performing any activity for which the member will receive any form of compensation from anyone other than the district.
13. Making any personal phone call that lasts longer than five minutes.
14. Sleeping except in designated areas and during prescribed times.
15. Watching television or engaging in other recreational activities except during prescribed times.
16. Remaining on duty for more than two continuous shifts unless ordered to do so by the fire chief during an emergency or disaster.
17. Permanently parking or storing vehicles, trailers, campers, tractors, boats, and so forth, on department property. Vehicles with commercial advertising on them shall be parked away from the station and shall not be readily visible to passers-by.
18. Making a false statement in any official communication or in conversation with another member or citizen.
19. Performing any act or making any statement, oral or written, about one's immediate superior, intending to destroy discipline and good order.
20. Performing any act or making any statement, oral or written, about one's coworkers, intending to destroy morale, good order, or working relationships with coworkers.
21. Displaying insolence or indifference or evading duty during an emergency incident. Any member found to be guilty of this offense shall be relieved of duty immediately.

Rules and Regulations, SOP 103.03

OUTSIDE EMPLOYMENT

1. Members who wish to accept part-time employment with another agency or organization in addition to their regular duties with the department must first obtain written authorization from the fire chief. (Local practice may vary.)
2. Employees shall forward such a request in writing to the fire chief. The request shall describe the work to be performed and the approximate number of hours per week that the employee wishes to work.
3. Outside employment shall not interfere with an employee's ability to satisfactorily perform his duties with the department.
4. Approval to work outside the department may be rescinded if an employee fails to satisfactorily perform his duties with the department.

Rules and Regulations, SOP 103.04

SEXUAL HARASSMENT

1. Unwanted or unsolicited verbal or physical harassment of members by supervisors or coworkers will not be tolerated. Supervisors shall promptly correct such behavior should it occur.
2. If a member informs a supervisor or coworker that his language or behavior is offensive and such conduct continues, the member immediately should report the situation to his supervisor or the supervisor's supervisor.

3. Appropriate disciplinary action shall be taken against a member found guilty of harassing a fellow member.
4. All members shall comply with the county's sexual harassment policy as described in the Human Resources Handbook.

Rules and Regulations, SOP 103.05

USE OF TOBACCO PRODUCTS

1. Buildings and structures owned and operated by the department have been designated tobacco-free workplaces. The use of tobacco products is not permitted inside a building or structure owned or operated by the department.
2. The use of tobacco products is prohibited by members while riding in or operating any vehicle owned or operated by the department.
3. While tobacco products may be used outside buildings and structures in smoking areas designated by the fire chief, members shall properly dispose of cigarette butts and other waste products.
4. Members using smokeless tobacco products shall refrain from spitting on sidewalks, on parking lots and other paved surfaces, on nonpaved surfaces used by other members, and in water fountains.

Rules and Regulations, SOP 103.06

EMPLOYEES SERVING AS VOLUNTEERS

Note: Rules concerning volunteer activity by paid employees vary from jurisdiction to jurisdication, but if your department is a combination department, this item should be addressed in your SOPs and rules and regulations. It would be wise to consult with your department's attorney when developing such rules.

1. Nonexempt employees may not join, serve, or be members of a county volunteer fire or rescue company in the same or similar capacity in which they serve as full-time employees. Employees may volunteer their services in other capacities. For example, a paid firefighter may not serve as a volunteer firefighter, but he may serve in an administrative capacity.
2. An employee may not serve as the president, fire chief, rescue chief, or member of the board of directors of a county volunteer fire or rescue company.
3. An employee may not serve as a volunteer company representative in other fire and rescue groups, commissions, or councils or on committees appointed by such organizations.

Section 200

General Administration

General Administration, SOP 200.01

STATION SUPPLIES

I. **Scope**

This standard regulates the procurement of consumable supplies. It was promulgated to ensure that the proper supplies will be available for cleaning and maintaining the department's apparatus, equipment, and buildings.

II. **Procedure for Ordering Supplies**

A. When supplies are needed, the member requesting them shall complete a Station Supply Requisition Form.

B. The member shall specify the type and quantity of each item being requested. The stock number or commodity code, if known, should be included for each item requested. The completed form shall be forwarded to the supply officer.

C. The supply officer shall compile the requisitions, order the supplies, and record the disposition of the order on the Station Supply Requisition Form. Supplies will be routinely delivered to the stations on ________ (day/date) of each ________ (week/month).

D. Occasionally, supplies must be obtained immediately so as to make repairs or complete an assignment. When supplies are obtained outside the normal procurement procedure, the member responsible for obtaining the supplies shall report their purchase to the supply officer as soon as practical by completing a copy of the Station Supply Requisition Form. The receipts for the items purchased shall be attached to the completed form.

III. **Responsibilities**

A. The fire chief shall:

1. Designate a captain or other member to serve as the supply officer.
2. Evaluate the performance of the supply officer on an annual basis.
3. Approve or deny any purchase of consumable supplies that exceeds $______ per item or a total cost of $______.

B. The supply officer shall:

1. Establish a minimum inventory of consumable supplies to be maintained at each fire station and ancillary facility.
2. Maintain an up-to-date ledger of expenditures for each station and ancillary facility operated by the department.

3. Approve or deny requests for purchases of consumable supplies that exceed $_____ per item or for multiple items that exceed a total cost of $_____.
4. Refer requests for supplies that exceed $______ to the fire chief or his designee for approval.
5. Notify the appropriate member in writing if an order doesn't meet the department's purchasing guidelines.
6. Provide each station with a current list of the supply inventory and the forms required to procure supplies.

C. Station officers shall:
1. Maintain and replenish the inventory of consumable supplies assigned to their respective fire station or ancillary facility.
2. Maintain a record of all purchases.
3. Order supplies on ______ of each week/month by completing an itemized Station Supply Requisition Form.
4. Regulate the use of consumable supplies and correct any misuse that may occur.
5. Report to the fire chief or his designee the theft of any items immediately on discovery of the loss.

General Administration, SOP 200.02

ON-DUTY MEALS

I. Scope

This standard regulates the procurement of groceries and meals by on-duty shift personnel.

II. General Guidelines

A. Ideally, meals should be planned prior to the beginning of each shift and groceries and other food items purchased prior to the beginning of the shift whenever possible. However, in the event that any of the on-duty personnel failed to prepare for the day's meals, the following guidelines shall apply:
1. Personnel may travel in their assigned apparatus to and from a grocery store or restaurant located within their respective district. They may *not* travel outside their district to obtain food unless they are returning from an assignment outside of their district.
2. The company shall remain in service at all times and properly park their vehicle in a designated parking space. At no time shall a vehicle park in a fire lane or other restricted space.
3. Crew integrity shall be maintained at all times. Only one member of the crew will be allowed to enter the grocery store or restaurant to purchase the food. That member shall take a portable radio with him and will remain in contact with the apparatus crew at all times. The other crew members will remain with their apparatus and shall maintain radio contact with Dispatch.
4. The member entering the store or restaurant shall be required to wear an appropriate uniform.

B. Exceptions:
  1. A company may stop at a convenience store to purchase refreshments when returning to the station from an alarm, inspection, training session, or other department-sanctioned activity. The store must be generally along the normal route of travel.
  2. Only one member will be allowed to enter the store and make the purchase. The other crew members will remain with the apparatus, and the provisions of Section A, above, will apply.
  3. It will be permissible for the member to wear a T-shirt in lieu of the uniform shirt when returning from an alarm or training session. At no time shall a member enter a business while in shorts.

C. Students and observers:

EMS students and other observers riding on an apparatus will be expected to comply with these rules.

III. Responsibilities

A. The company officer shall be responsible for strictly enforcing the provisions of this procedure.

B. Each member will be expected to strictly observe the provisions of this procedure and will not be excused in the event that an officer is not present.

General Administration, SOP 200.03

LOST/DESTROYED EQUIPMENT

I. Scope

This standard sets forth the requirements for reporting the loss or destruction of equipment owned or operated by the department.

II. Reporting Procedure

A. On discovery that a piece of equipment has been lost, damaged, or destroyed, an employee shall record his findings on the Report of Equipment Lost or Destroyed form and forward it through the chain of command to the fire chief.

B. If the lost or damaged item is replaced from existing inventory, note it on Line 9 of the report.

C. The asset numbers and replacement costs should be recorded for budget and inventory purposes.

D. For training or disciplinary purposes, complete Lines 7 and 8.

III. Responsibilities

A. It is the responsibility of each officer to maintain all equipment assigned to his station and apparatus in a constant state of readiness. To facilitate this process, each apparatus is to be inspected and inventoried at the beginning of each shift and after each incident during the process of returning the apparatus to service.

B. It is the responsibility of every member of the department to properly use and maintain the equipment assigned to him.

IV. Accountability

A. The deliberate or willful misuse, theft, loss, damage, or destruction of any tool, equipment, or other device owned by the department or other agency or private individual will result in appropriate disciplinary action as prescribed in this department's rules and regulations manual.

B. As a part of the disciplinary process, the individuals responsible for the loss or destruction may be required to reimburse the department for the costs to repair or replace the equipment.

General Administration, SOP 200.04

FIRE DEPARTMENT LIBRARY

I. **Scope**

This standard regulates the use of printed and audiovisual materials owned by the department. It was promulgated to establish a department library consisting of printed and audiovisual materials. These materials are to be used to increase public awareness of fire safety, medical emergencies, and disaster preparedness issues. In addition, the library should be used to increase the collective knowledge of the members of the department.

II. **Station Library**

A. Each fire station shall maintain a small library for use by on-duty personnel. The library materials will consist of periodicals, a copy of the adopted fire and building code, at least one copy of study materials for promotional examinations, and any other relevant materials.

B. These materials may not be checked out for individual use, loaned to another station, or removed from the fire station without the written permission of the fire chief.

III. **Department Library**

A. The department or general library shall be maintained in the administrative/training facility of the department. This library shall house the various audiovisual materials and equipment used in training, as well as periodicals, study materials, fire and building codes, and printed materials.

B. Generally, most of these materials will be available for individual use and checkout except where otherwise prohibited. At least one set of all study materials shall be placed on reserve and shall not be removed from the library.

C. From time to time, the department may issue printed items or textbooks to individuals for their use while employed by the department. The department retains continuous ownership of these materials. When these items are issued to an individual, the title of the item and the date of issue shall be recorded and placed in the individual's permanent personnel file.

IV. **Check-Out and Return Procedure**

A. An individual may borrow books from the department library for a period not to exceed 14 days. The borrowed item must then be returned to the library. If no one has requested to borrow the book, it may be checked out for another 14 days.

B. No individual will be allowed to check out more than two books at any one time.

C. To check out a book, an individual must sign and date the book card located inside the front cover of the book. The book card will then be retained in the library card file.

D. The individual shall also record the item being borrowed on the Library Book Check-Out Log.

E. Books may not be loaned to another individual while checked out. The person who originally borrowed the book will remain responsible for it.

F. Audiovisual materials and equipment may also be checked out by a member of the department, another district employee, or a neighboring fire department. The individual shall complete an Audiovisual Materials Check-Out Form.

G. When an item is returned to the library, the borrower shall retrieve the book card and record the return date and again place the book card inside the front cover. The return date shall also be recorded on the Library Book Check-Out Log.

V. **Responsibilities**

A. The officers assigned to each fire station shall be responsible for maintaining their respective station libraries and for regulating the use of its resources.

B. The deputy fire chief and the fire marshal shall be the custodians of the department library. They shall maintain the materials within the library and regulate the use of its resources.

C. Any individual who borrows or checks out materials from the department library shall be responsible for the materials entrusted to him. He shall be required to replace any item that is damaged or destroyed while in his care.

D. Any individual who leaves the employment of the department shall be responsible for returning all books and other printed material that has been issued to him. He shall be required to replace any item that has been lost or damaged.

General Administration, SOP 200.05

DAILY WORK SCHEDULE

I. **Scope**

This standard regulates the daily activities of personnel assigned to fire station duty.

II. **Daily Schedule**

A. The following tasks shall be performed during every shift, as incident volume and weather permit:

Time:	Daily Activity:
07:00	Staffing officers report for duty.
08:00	Shift change, roll call, raise the flags, and announcements.
08:15	Check apparatus and equipment, clean up the station, and police the grounds.
09:00-11:30	Company inspections, training, physical fitness, other scheduled activities.
11:30-13:00	Lunch.
13:00-17:00	Per 09:00 to 11:30.
After 17:00	Unstructured activity, study, and incident response.

Flags are to be lowered prior to nightfall.

B. Breaks may be taken as time and activities permit.

C. Watching television is permitted during the lunch period and after 17:00 hours. Watching the Weather Channel, televised training, and relevant videotapes is permissible anytime during the tour of duty.

D. Sleeping is permitted after 21:00 hours and during the lunch break. Abnormally high incident volume may also necessitate additional rest periods.

E. With the permission of the company officer, physical fitness activities may be conducted prior to 17:00 hours at a fire station or city/county-owned facility that is within a company's first response district. After 17:00 hours, companies may use church, school, or university facilities within their first response district. Companies may also swap districts, if necessary, so as to complete their required fitness activities.

III. Assigned Tasks

A. In addition to routine apparatus checks and general housekeeping activities, the following maintenance tasks are to be performed as incident volume and weather permit:

Day:	Task:
Sunday	Wash and service reserve apparatus, clean dorms and bathrooms, wax floors, conduct training.
Monday	Clean and inspect in-service apparatus; clean apparatus bays, shops, and tool rooms; inventory EMS supplies; and wash automobiles assigned to the fire chief and fire marshal.
Tuesday	Truck day. Set up and raise the aerial device and test all operations, inspect all ground ladders, wash automobiles assigned to the deputy chief and public education officer.
Wednesday	Clean and wax all kitchens and day rooms, clean refrigerators, wash automobiles assigned to the fire protection engineer and fire investigator.
Thursday	Pump day. Wash reserve automobiles and pickups, check and service station emergency power supplies; check and service rehab and cascade unit.
Friday	Conduct training; clean and wax all offices, hallways, and foyers; wash maintenance pickup; check and service SCBA compressors.
Saturday	Lawn and grounds day. Clean windows, conduct training.

B. The daily schedule shall be suspended on Thanksgiving and Christmas Day. The only activities that must be performed are those activities required to maintain apparatus and equipment in a state of readiness.

IV. Responsibilities

A. The company officer shall be responsible for ensuring that all assigned tasks are completed each shift if incident volume and weather permit.

B. The company officer, at his discretion, may also alter the daily schedule provided that all assigned tasks are completed before the end of each shift.

General Administration, SOP 201.01

MINIMUM STAFFING

I. Scope

This standard regulates the daily, routine staffing of fire stations and apparatus by sworn personnel. The provisions of this procedure may be suspended by the fire chief or his designee whenever special circumstances warrant. This standard was promulgated to:

A. Establish guidelines that are intended to provide the community with the highest quality fire and EMS service possible within the parameters of the department's budget.

B. Minimize the health and safety risk of personnel by assembling a sufficient number of personnel at every incident to bring the incident to a safe and satisfactory conclusion.

II. Staffing Guidelines

A. A minimum of _____ personnel shall be on duty at all times. Any deviation from this standard must be approved by the fire chief or, in the chief's absence, his designee.

B. To ensure compliance with this standard, the fire chief shall appoint a staffing officer for each shift. The shift staffing officer shall be responsible for maintaining staffing levels at prescribed minimums. To fulfill this responsibility, the shift staffing officers shall have the authority to assign, move, or transfer personnel as necessary.

C. The shift staffing officer shall be responsible for staffing a minimum of ____ engine companies, ____ truck companies, and ____ medic units. These companies will be deployed in ____ fire stations.

D. Personnel shall be assigned as follows:

1. Medic units: Minimum staffing shall be ____ paramedics. This does not include student observers who may be assigned to ride with the unit.
2. Engine companies: Minimum staffing shall be __ personnel: an officer, a driver, and __ firefighters. At least __ crew member(s) shall be a paramedic in those companies assigned to stations without a staffed medic unit.
3. Truck companies: Minimum staffing shall be __ personnel: an officer, a driver, and __ firefighters. At least __ crew member(s) shall be a paramedic.

E. Whenever fewer than ___ personnel report for duty, a sufficient number of off-duty personnel shall be hired on overtime to satisfy the minimum staffing requirements. (For volunteer departments with assigned duty crews, members from off-duty crews shall be contacted to fill the vacant slots.)

F. When more than ___ personnel report for duty and all minimum staffing requirements have been met, any extra personnel that remain on duty may be allowed to remain with their regularly assigned companies.

G. Emergency transfers to a medical facility outside of the department's EMS district shall be handled by off-duty personnel whenever possible. On notification of the need for a transfer team, Dispatch shall page off-duty personnel and request a two-member transfer team. The team must be composed of at least __ paramedic(s) who shall attend the patient. The driver does not need to be a paramedic.

III. **Responsibilities:**

A. The shift staffing officer shall:

1. Be accountable for the location and duty status of all personnel assigned to his respective shift. This shall include all personnel who may be temporarily assigned to his shift because of overtime, shift swaps, etc.
2. Prepare a daily staffing report and forward it to the fire chief at the conclusion of each shift.
3. Generate an overtime roster for his respective shift. Personnel shall be placed on the roster according to their rank and in order of seniority within their respective rank. Whenever possible, overtime assignments shall be given to members of the off-going shift.
4. Determine daily the number of personnel that must be hired for each shift. A minimum of ___ personnel to meet operational requirements shall be held over at the beginning of each shift.
5. Notify standby personnel that they may be needed for overtime and that they are being placed on standby status. This should be done as soon as practical.
6. Maintain an accurate record of each employee's attendance. This should include overtime worked and any leave time used. This information should be recorded on the daily report and forwarded to administration.
7. Approve or disapprove all types of leave requests. Exception: Shift swaps between members of different ranks may only be approved in unusual circumstances and only by authorization of the fire chief.

B. Dispatch shall:

1. Make the following radio announcement on notification that the daily staffing assignment is complete: "Staffing is complete. Standby personnel are released."
2. Make the following radio announcement whenever operational conditions preclude completing staffing assignments prior to shift change: "Shift staffing is not complete. Standby personnel need to remain on duty."
3. Make subsequent announcements as requested by the shift staffing officer.

C. Sworn personnel shall:

1. Not leave their assigned duty post until they are properly relieved. If their relief does not report for duty, the staffing officer shall be notified.
2. Apply to their shift staffing officer so as to be added to the overtime roster and contact the staffing officer if they desire to be deleted from the roster.
3. Notify the staffing officer of their availability to work overtime. If placed on standby status, no compensation shall be paid unless they are hired.

4. Be rotated to the bottom of the overtime roster, plus one additional complete rotation of the roster, for refusal to work overtime. (This practice will vary by jurisdiction and is only one of many available methods.) Exceptions: The staffing officer may lessen this penalty whenever pressing operational requirements make it necessary to do so, and he may waive the penalty in the event of a verifiable personal or family emergency.
5. Any member who leaves an assigned duty station prior to Dispatch's announcement that staffing is complete shall be rotated to the bottom of the overtime roster. Anyone guilty of this infraction shall be ineligible for hire for the remainder of the existing roster, plus one additional complete rotation of the roster.
6. Be ineligible for working an additional overtime assignment in the event that he has worked more than __ hours overtime during the life of the current overtime roster.

IV. <u>Method of Filling Vacancies</u>

A. Whenever a vacancy exists, the appropriate person or persons shall be hired from the overtime roster. Personnel shall be hired as follows:
  1. Vacancies shall be filled by rank, certification or training, and seniority in accordance with department policies.
  2. Whenever a vacancy occurs after Dispatch has announced that staffing has been completed and the standby personnel have been released, the staffing officer shall contact employees on the overtime roster to fill the vacancy.
  3. Anyone passed over for overtime due to duty classification shall retain his position on the overtime roster.
  4. A member who is sick, is on injury leave, is serving a disciplinary suspension, or has worked more than ___ consecutive hours shall not be hired for an overtime position. The member shall retain his position on the overtime roster until he returns to duty. Working overtime is optional for members on vacation leave.
  5. A member is entitled to an unlimited number of free passes for going on the standby list without affecting his position on the overtime roster. Members on the standby list shall be hired first.

B. Personnel hired on overtime shall report to their duty post on the appropriate date and at the appropriate time and shall report with all the required uniforms, protective clothing, bed linens, and personal items appropriate for their assignment. Anyone who reports without these items shall be released from duty and rotated to the bottom of the overtime roster.

C. Members shall notify the appropriate supervisor if they cannot report as assigned and shall provide the supervisor with a satisfactory explanation and their estimated time of arrival.

D. A member working an overtime assignment may, with the prior approval of both his supervisor and the staffing officer, find another employee to work the balance of his assignment. His relief *must* be:
  1. Eligible to work the assignment.
  2. Paid by the department and not the employee.
  3. Paid during the payroll that the hours were actually worked.

E. If a member secures a relief from an overtime assignment, he shall still be considered as having worked his full assignment for consideration for future overtime assignments.

V. **Emergency Callback**

A. Whenever operational conditions are such that additional personnel are required, one or more off-duty personnel may be called back to duty.

B. Emergency callbacks shall be initiated as follows:

1. On the request of a ______________ (insert second, third, etc) alarm (fire or medical) by an incident commander, all off-duty personnel shall be paged by Dispatch and requested to return to duty. Off-duty personnel shall be instructed to either staff reserve apparatus or respond directly to the incident.
2. During periods of high activity, off-duty personnel shall be paged to return to duty to staff reserve apparatus whenever the number of in-service companies is reduced to as few as _______ engine companies and _____ medic unit(s).

C. Unless otherwise advised, all personnel answering an emergency callback shall respond to their assigned station. Off-duty personnel shall notify Dispatch of their availability. Dispatch shall relay the information to the incident commander. It may be necessary to move personnel to other stations to properly staff reserve apparatus.

D. All off-duty personnel requested to respond directly to an incident shall report to the staging officer for assignment. At no time shall a member begin any task without authorization from either the staging officer or the incident commander.

E. The issue of compensation should be discussed for career employees. The practice will vary by jurisdiction and must take into consideration FLSA provisions, state and local civil service laws, and bargaining agreements. Typically, emergency callback duty will require the local jurisdiction to compensate the employee at a rate of 1.5 times his normal base rate of pay. Some jursidictions allow the employee to accept compensatory time in lieu of overtime. Check with your attorney and human resources department.

General Administration, SOP 201.02

COLLATERAL ASSIGNMENTS

I. **Scope**

This standard regulates the assignment of collateral duties to selected personnel and authorizes assignment pay for the performance of certain duties that exceed the normal scope of a given job classification. This standard was promulgated to:

A. Provide management and leadership training to the members of the department.

B. Limit the number of full-time administrative personnel by delegating administrative and logistical support functions to shift personnel.

II. **Assignment Criteria**

A. The assignment of collateral duties and assignment pay slots will be based on:

1. The necessity or desirability that the activity or assignment be performed.
2. The ability of the department to fund the activity.
3. A member's ability to successfully perform the assigned task or to acquire the new skills that may be required to accomplish the task.
4. A member's performance of previously assigned duties and tasks.
5. The willingness of a member to perform the task.

B. Assignment procedure:

1. During ____________ of each fiscal year, the fire chief will assign collateral duties to each company officer. The term of the assignment will coincide with the fiscal year.
2. The assignments will be rotated as often as practical so as to improve the overall skill levels of the program's participants.

III. Collateral Assignments

A. The following collateral positions will be assigned by the fire chief:

1. Communications officer: Serves as a liaison with the communications divisions of other departments and shall be responsible for inventorying and maintaining all two-way radios, pagers, and cellular telephones owned by the department, as well for updating the communications SOPs.
2. Facilities and energy management officer: Manages the distribution of all consumable station supplies and coordinates the repair of the department's buildings and facilities.
3. Haz-mat officer: Coordinates all of the haz-mat training required by state and federal agencies and is responsible for tier-two reporting, representing the department at the local emergency planning committee meetings, haz-mat SOPs, and the maintenance of equipment and supplies used to control haz-mat incidents.
4. Hose, ladder, and small-equipment officer: Responsible for the testing, inventory, and repair of all fire hose, ground ladders, fire extinguishers, and other small equipment owned by the department, excluding communications equipment, protective clothing, SCBAs, and medical equipment. He is responsible for maintaining the applicable SOPs and completing the associated reporting functions.
5. Management information system officer: Responsible for the maintenance and use of all data processing and computer equipment owned or operated by the department and for the collection and analysis of data as may be required by the fire chief.
6. Maps/street index officer: Prepares, maintains, and updates all maps used by the department and the master street index used by Dispatch.
7. Recruitment/retention officer: Responsible for the recruitment of new members and assisting with the retention, development, and promotion of existing members.
8. Safety officer: Responsible for safety and health-related SOPs; compliance with NFPA 1500, *Standard on Fire Department Occupational Safety and Health Program;* fireground accountability; and scheduling monthly safety meetings and serves as chairman of the Driver's Accident Review Board.

9. SCBA officer: Responsible for the maintenance and repair of all compressors, cascades, breathing apparatus, and related equipment. Also responsible for the required testing, record keeping, reports, and related SOPs.
10. Shift training officer: One per shift. Refer to SOP 202.01, Shift Training Officers.
11. Staffing officer: One per shift. Responsible for maintaining minimum staffing levels; the assignment of personnel to stations, fire companies, and medic units; and the coordination of vacations, overtime, etc. Refer to SOP 201.01, Minimum Staffing.
12. Uniform and protective clothing officer: Prepares specifications for and purchases all uniforms and items of protective clothing except SCBA. Maintains an inventory of uniforms and protective clothing and issues items as required. Responsible for budget preparation and compliance with NFPA and state OSHA regulations.
13. Water supply and tactical survey officer: Serves as a liaison with water utilities, maintains and updates hydrant maps and fire-flow information, and identifies and maps static water supplies. Coordinates the hydrant testing and marking program and is responsible for the assignment and development of tactical surveys and ensuring that each apparatus carries the same.
14. Wellness officer: Maintains all wellness equipment, chairs the Wellness Committee, supervises the mandatory physical fitness program, serves as a liaison to the district's Wellness Committee, coordinates the annual medical and physical ability exams, and serves as the EAP officer.

IV. **Assignment Pay**

A. Assignment pay may be authorized by the city council for certain positions.

B. The compensation for each position is set during the annual budgeting process.

V. **Responsibilities**

Each person assigned a collateral duty or an assignment pay position is responsible for performing all of the related tasks and shall be accountable to the fire chief for his performance. Each individual so assigned shall prepare a monthly report summarizing his activities and forward a copy to the fire chief by the fifth day of each month.

General Administration, SOP 201.03

COMPLAINTS AGAINST EMPLOYEES

I. **Scope**

This standard establishes guidelines for the receipt, investigation, and resolution of complaints received by the department concerning the professional or personal conduct, behavior, action, or inaction of one or more members and those complaints that concern the department as a whole. It was promulgated to ensure that a thorough investigation be conducted for every complaint received by the department; that the innocent be exonerated and the guilty be properly punished.

II. General

A. Complaints that arise from the daily conduct of business fall into one of two major categories:

1. Complaints against the department as a whole; and
2. Complaints against one or more individual members.

B. Complaints generally allege a violation of a departmental rule, policy, procedure, or general order. A complaint may also allege that there has been a violation of a federal, state, or local statute or ordinance.

C. The department will hear all complaints against its members that have been initiated by any person who is found to have standing for such a complaint. Persons other than those who are actually affected by the actions of a member shall have no standing for a complaint, with the exception of cases that involve juveniles. In such cases, the parent or legal guardian shall be required to file the written complaint.

III. Complaint Procedure

A. Complaints must be made in writing and shall be signed by the persons making the complaint. The complaint need not be in affidavit form but should be filed on a Complaint Form.

B. Complaints may be accepted by any on-duty supervisor. The supervisor who receives the complaint shall conduct an initial investigation to obtain as much information as possible and then forward the complaint directly to the fire chief in the most timely manner possible.

C. A signed letter of complaint may be accepted in lieu of the Complaint Form if the supervisor has verified that the letter is not fictitious and is able to verify the name of the complainant.

D. If the complainant does not want to file a written complaint or does not wish to sign the Complaint Form, he may still voice his complaint to a supervisor. The supervisor shall record the information and forward an account of the complaint or the unsigned form to the fire chief.

E. On receipt and review of an oral or written complaint, the fire chief shall assign the investigation of the complaint to an internal affairs investigator. (Some departments use a chief officer or local law enforcement agency for this purpose.)

F. An investigator may also be sent to a complainant's home or place of business, if necessary or so requested, to accept a formal written complaint or to obtain additional information.

G. Where there are mitigating circumstances, a complaint may be initiated by telephone. A tape recording of the initial complaint shall be made, if possible, and should be retained by the supervisor. The complaining party should be asked to submit his complaint in written form as soon as possible.

H. A copy of the written complaint shall be given to the accused member at the time that the member is requested to make a written reply to the allegations.

I. When the act described in the complaint is a crime, the circumstances will be immediately explained to the fire chief. The fire chief should then determine if the accused member should be arrested forthwith, if a warrant should be obtained for his arrest, or if there is a need for further investigation before any action is taken. He should also determine whether or not the member should be suspended pending the outcome of the investigation.

J. When there are indications that the member cited in the complaint may have been or is under the influence of a controlled substance, he should be asked to submit to a substance screening test. The lapse of time, expressed in minutes, between the initial report or observations of the member's conditions and when the test is administered should be accurately recorded. The accused member should not be allowed to drive but shall be transported to the test site by a supervisor.

IV. Resolution of the Complaint

A. On conclusion of an investigation, the complaint shall be classified as one of the following:

1. Unfounded: The allegation has been proved false or there is a lack factual evidence to support it.
2. Exonerated: The incidents cited did occur, but the actions were lawful or followed proper procedures.
3. Not sustained: There is insufficient evidence either to prove or disprove the allegations.
4. Sustained: The allegations are supported by sufficient evidence and the complaint will be upheld.

B. Resulting Action:

1. When an investigation results in a determination of *unfounded, exonerated,* or *not sustained,* the fire chief will notify the accused member that no further action will be taken against him. The member shall return to duty. If he had been placed on unpaid leave, he shall be fully compensated for all lost wages and benefits.
2. When an investigation results in a determination that an allegation is supported by sufficient evidence, the appropriate action will be taken in accordance with the department's disciplinary policy.

V. Format

A. The supervisor who accepts the original complaint or conducts the initial investigation shall make his report as complete as possible prior to submitting it to the fire chief.

B. All internal affairs investigators shall use the following format for conducting their investigations:

1. Record the facts surrounding the incident. What took place when and where, and who was involved? Be brief.
2. State the allegation in detail. What does the complaint allege and against whom?
3. Record the process used to investigate the complaint. What did you do as the investigator? What did you learn from talking to all the parties and witnesses? List them.
4. Record all findings and the conclusion reached. What did your investigation reveal based on the facts extracted from the evidence?
5. Include the written recommendation to the fire chief.

VI. Exceptions

A. A complaint should be referred to the fire chief or other senior staff member in those cases where the complainant prefers to speak *only* to those entities.

B. Internal investigations may also be conducted concerning a member's conduct whenever the fire chief has reason to believe doing so is warranted.

C. The procedure in Section III, above, may be waived if the fire chief determines that mitigating circumstances warrant such a waiver.

D. The fire chief shall have absolute and final authority in determining whether a disciplinary action should be taken.

General Administration, SOP 201.04

EVALUATION OF SWORN PERSONNEL

I. Scope

This standard applies to the members of the department who are subject to the provisions of (insert enabling legislation) ______________________.
This standard was promulgated to:

A. Formally communicate the goals and objectives of the department to each member and to discuss the member's individual role in the accomplishment of those goals and objectives.

B. Improve the performance and productivity of each member.

C. Identify each member's need for additional training and education.

D. Document in writing each member's performance and identify corrective actions that a member might be required to make so as to improve his performance.

II. Evaluation Process

A. A formal, written evaluation of each member's performance shall be conducted by his immediate supervisor once each year within 30 days of the member's anniversary date.

B. For guidance in conducting the evaluation, the supervisor should consult *Employee Performance Evaluations: A How-To Process for Fire and Rescue Service Leaders* by James E. Gerspach (Ashland, Massachusetts: International Society of Fire Service Instructors, 1988).

C. The evaluation will be recorded on the department's Annual Employee Performance Review Form.

D. During the formal evaluation process, each member shall be counseled by his immediate supervisor with respect to his individual progress and development. The supervisor shall note any area of concern and shall discuss the steps that the member should take to correct the problem or deficiency. A summary of the discussion shall be recorded on the member's evaluation form.

E. On completion of the initial evaluation session, the evaluating supervisor shall forward the evaluation form to the appropriate supervisor for his review.

F. The reviewing supervisor shall discuss the member's progress and deficiencies with the evaluating supervisor. If there is agreement about the evaluation, the reviewing supervisor shall add his comments and sign the form. If there is disagreement, the two supervisors shall record the changes that need to be made on the evaluation form. The reviewing supervisor shall add his comments, sign the form, and return it to the member for his review.

G. A member may disagree with any portion of the evaluation and shall be allowed to record his objections in the section reserved for member comments.

H. On completion of the review process, the member shall be required to sign the form. His signature does not imply agreement—only that he has read the evaluation form and has been made aware of the contents of the document. Under no circumstances shall a member be required to sign a blank or incomplete evaluation form.

I. Each member shall be furnished a copy of the completed evaluation form. The original, signed copy will be placed in his permanent personnel file.

J. No changes will be made to the form after all parties have signed off on the evaluation unless all parties are informed of the changes and are furnished copies of the changes.

III. Responsibilities

A. Supervisors shall be responsible for completing a formal written and oral evaluation of the performance of each member under their supervision at least once each year in accordance with the provisions of this standard.

B. Members shall be responsible for participating in a formal written and oral evaluation of their performance by their supervisor and for correcting all deficiencies identified in the evaluation process.

C. The fire chief shall be responsible for ensuring that copies of each member's performance evaluations are maintained in their permanent personnel files.

General Administration, SOP 202.01

SHIFT TRAINING OFFICERS

I. Scope

This standard outlines the duties and responsibilities of officers and other members assigned the collateral duty of shift training officer.

II. Responsibilities

The shift training officer shall be responsible for:

A. Maintaining an up-to-date training file of each member assigned to his shift.

B. Forwarding to the fire chief by the fifth day of each month a report of the shift's training activities for the previous month.

C. Developing an annual program of basic training in conjunction with the other shift training officers and supervising the delivery of the program on his shift.

D. Administering a territory test to each shift member once each month. The territory test shall test each shift member's knowledge of the location of streets, fire hydrants and static water supplies, standpipe and sprinkler connections, and target hazards.

E. Scheduling and conducting shift training sessions.

F. Monitoring and tracking each shift member's compliance with fire and EMS minimum continuing education requirements as promulgated by OSHA and other appropriate agencies.

G. Assisting with the professional development of each shift member.

H. Assisting with the development and dissemination of training materials.
I. Supervising the orientation and progress of probationary personnel assigned to his shift and making recommendations concerning the retention or dismissal of probationary personnel.
J. Other such duties as may be assigned.

General Administration, SOP 203.01

TRAINING/TRAVEL REQUESTS

I. **Scope**

This standard outlines the process that must be followed when a member wishes to be compensated for attending a seminar or training session that isn't being hosted or sponsored by the department. It was promulgated to enable members to attend training classes and seminars conducted by outside agencies and organizations.

II. **Procedure**

A. If a member wishes to obtain approval to travel on behalf of the department or to attend a training class or seminar, the member must complete a Request for Travel/Training Form at least 30 days prior to the date of the class or seminar and submit the completed form to his immediate supervisor.
B. If the request is from a member assigned to shift work, the approved form shall then be forwarded to the shift staffing officer. The staffing officer will record the impact the request would have on minimum staffing on the request form and forward the request to the shift commander.
C. If the shift commander approves the request, the approved form shall be forwarded to the fire chief for his approval or disapproval.
D. If the training request is from a member who isn't assigned to shift work, the approved request shall be forwarded to the appropriate division commander for his approval.
E. If the division commander approves the request, the approved form shall be forwarded to the fire chief for his approval or disapproval.
F. If the request from a member assigned to a shift is approved by the fire chief, the shift staffing officer shall be notified and the necessary slots shall be reserved on the shift leave calendar.
G. If a request is disapproved at any point in the process, the person disapproving the request will record his reason for disapproval on the form and return the request to the member who submitted the request.
H. As a general rule, any request that requires the payment of overtime must be approved by the fire chief. Budget constraints and the value of the class to the department shall be considered for all requests.

General Administration, SOP 203.02

COMPENSATION/REIMBURSEMENT FOR TRAVEL

I. **Scope**

This standard establishes guidelines for reimbursement and compensation of members who are required to travel or attend meetings, training sessions, or seminars on behalf of the department. It was promulgated to ensure that the department's compensation and reimbursement procedures are in compliance with the Fair Labor Standards Act and the applicable district policies and procedures.

II. **Criteria for Reimbursement of Expenses**

A. To be reimbursed for expenses incurred while traveling or attending training sessions on behalf of the department, personnel must:

1. Obtain the fire chief's approval to travel or attend a meeting, training session, or seminar.
2. Submit an itemized expense report with receipts and proper documentation.

B. The department will normally cover the following expenses:

1. Tuition or registration and course materials.
2. Accommodations at an approved motel or hotel.
3. Business-related long-distance calls, plus one nonemergency personal call home per day. The duration of calls shall be limited.
4. Parking and ground transportation based on receipts and actual expenses. Prior permission to rent an automobile must be obtained before reimbursement will allowed.
5. $_________ per day maximum allowance for meals and tips.
6. For local travel, the department will furnish a vehicle whenever possible. Members using their personal vehicles will be reimbursed according to district policy.
7. Members will be expected to travel by air whenever it is practical to do so. Reimbursement will be based on round-trip coach fares from the airport. If a member elects to drive in lieu of flying, reimbursement will be limited to the normal airfare.
8. The department will not reimburse any expense incurred for alcohol or entertainment.
9. The department will not reimburse any expenses incurred by a member's spouse or other family members.

C. All unexpended funds paid to the member in advance will be returned to the district with the member's itemized expense report.

III. **Compensation for Travel**

A. Travel time to and from an event shall be compensable. Compensation will be based on the amount of time required to travel to and from the department's headquarters.

B. Nonexempt employees shall be granted overtime or compensatory time, as mutually agreed on prior to the event, for those hours worked in excess of their normal work week (40 or 56 hours). Compensable time includes travel time and class attendance. It does *not* include weekends or those evening hours after the event has concluded for the day or week.

C. Whenever possible, shift personnel will be assigned to a 40-hour week while attending an event with a duration of 40 or more hours.

D. Operations personnel will normally be granted a minimum of 12 hours off prior to leaving for a class and 24 off prior to returning to duty.

General Administration, SOP 203.03

RECORDS AND FORMS

I. **Scope**

This management procedure establishes guidelines for the collection and retention of information. It was promulgated to:

A. Provide for the collection and maintenance of information for legal record-keeping purposes.

B. Provide a database for the analysis of the activities of the department.

II. **Creation and Retention of Records**

A. This standard hereby creates a data collection system for the department.

B. The data collection system is subdivided into the following general categories:

1. Administrative files.
2. Criminal records and case files.
3. Inspection files.
4. Permanent occupational safety and health files.
5. Permanent personnel files.
6. Training records.
7. Vehicle and equipment maintenance and repair records.

C. Records and reports shall be retained in accordance with the applicable state law, district ordinance, and department policy.

D. No record or report shall be disposed of without the approval of the appropriate supervisor.

III. **Forms**

A. The fire chief shall cause the creation of such forms as may be deemed necessary to document the activities of the department. A new form may be proposed by any member of the department but shall not be published or distributed without the authorization of the fire chief.

B. All forms shall be assigned a permanent form number, and an inventory of form numbers shall be maintained by the fire chief's secretary. The form number shall be displayed in the upper left corner of each form, and the month and year that the form was created or revised shall be displayed in the upper right corner.

C. Each form shall be reviewed periodically to ensure that it is still necessary, and it shall be revised or deleted as deemed appropriate by the fire chief.

IV. **Responsibilities**

A. The fire chief shall be the custodian of the records and shall be responsible for:

1. Causing the creation and maintenance of those records and reports that are required for legal purposes.
2. Causing the creation and maintenance of the various reports and forms necessary to document the various activities of the department.
3. Ensuring the confidentiality of those records where required by statute or otherwise deemed appropriate.

4. Furnishing those records to the public that are properly requested under the state's Freedom of Information Act in accordance with the provisions of the statute.

B. The fire marshal shall be the custodian of all criminal records and case files generated by the department and shall be responsible for:
   1. Causing the creation and maintenance of those records and reports that are required by statute that pertain to the investigation and prosecution of those crimes and offenses against persons and property as addressed by the State/Local Government Code.
   2. Causing the creation and maintenance of those records and reports that are required by the fire chief for the prevention of fire and the enforcement of applicable sections of the fire and building codes.

C. Each employee shall be responsible for:
   1. Completing the forms and filing the reports required by the provisions of the Standard Operating Procedures Manual and the Rules and Regulations Manual. All forms and reports shall be complete and accurate.
   2. Filing and maintaining records and reports as required by departmental policy.
   3. Maintaining the confidentiality of departmental records and reports. Records and reports shall not be released without authorization by the appropriate supervisor.

General Administration, SOP 203.04

INCIDENT REPORTS

I. <u>Scope</u>

This standard establishes requirements for the preparation of reports for incidents to which the department responds. It was promulgated to:

A. Create a permanent record of each incident to which the department responds.

B. Develop a database for the analysis of the community's demand for fire and emergency medical services.

C. Provide uniform data to the state fire marshal and the NFPA concerning the department's emergency response activity.

II. <u>Completing the Report</u>

A. The department uses the incident reporting system promulgated by the state fire marshal.

B. A Basic Incident Report shall be completed for each incident to which the department is dispatched, and a copy of the report shall be forwarded by the department to the state fire marshal. If possible, the report shall be filed electronically using a suitable software program. If this is not possible, a report may be filed manually by completing a hard copy of the Basic Incident Report. A copy of the instructions for completing a Basic Incident Report has been provided to each fire station.

C. The report shall be accurate and thorough, and it shall contain sufficient information to allow the reader to re-create an accurate portrayal of the facts and events surrounding a given incident.

D. The dispatch office assigns a chronological number for each incident, and the member filing the report shall record the appropriate incident number on the Basic Incident Report.

E. The Basic Incident Report has been modified slightly to accommodate local needs. These changes are as follows:

1. Line E—Method of Alarm to Fire Department: All automatic alarms are reported to fire alarm telephonically. Alarms initiated by an alarm company or by on-site security shall be coded as a 3—Private Alarm System.
2. Line E—Type of Situation Found:
   a. An EMS incident that requires a Basic EMS Report be completed shall be coded 32.
   b. The type of situation found on arrival of the first company should be recorded on all other incidents.
   c. Fire companies that initiate a call for a medic unit after arriving at an incident to which a medic has not been dispatched on the initial assignment shall also enter the conditions found on arrival.
   d. If a medic unit is needed, a Basic Casualty Report shall also be completed.
3. Line F—District: District numbers are entered for the geographic location in which the incident occurs and coincide with the number of the first-due engine company. For example: 001 would be entered for an alarm in Engine 1's first-due area.
4. Line F—Outside Fire Service Assistance: Mutual aid given does not include responses to areas under contract with the city or county for fire or EMS service.
5. Line H—Number of Fire Suppression/EMS/Other Fire Service Apparatus and Personnel: Enter the actual number of personnel and apparatus dispatched to the incident, including those that may be disregarded or that respond on subsequent alarms.

F. Blank Basic Incident Report forms shall be carried on all in-service apparatus and medic units so that information may be gathered at the incident scene.

G. Whenever an incident is investigated by a member of the fire marshal's office or the arson task force, the fire investigator should be contacted by the member completing the Basic Incident Report to report the appropriate information concerning cause, point of origin, and estimated dollar loss.

H. A narrative shall be written for each incident. The narrative shall include a brief description of the events that occurred and the actions that were taken during the resolution of the incident. The narrative shall be thorough, concise, and accurate. Neatness and spelling count. The narrative shall be limited to the facts and *not* include superfluous or editorial comments.

I. The proper number and types of vehicles, the number and names of personnel at the scene, and the equipment that is used shall be entered on each incident report. The station number of the vehicle that arrives on the scene first shall be entered on the report. If a reserve apparatus is being used, the number of the vehicle that has been replaced shall be used and not the number of a reserve. Exception: The number of a reserve shall be used if the reserve has been placed in service by off-duty personnel due to multiple incidents or a large event.

J. A Basic Casualty Report shall be completed in addition to the Basic Incident Report if a person is injured or killed as a result of a fire.

K. The Basic EMS Report shall be completed for EMS incidents in lieu of the Basic Casualty Report.

III. **Responsibilities**

A. The officer of the first-arriving engine or truck company shall be responsible for the completion of the Basic Incident Report and all other reports that may be subsequently required. Exception: If there is no officer on the first-arriving apparatus or for EMS incidents, the highest-ranking or senior member of the crew shall be responsible for completing the Basic Incident Report.

B. The watch commander shall be responsible for reviewing the completeness and accuracy of all incident, casualty, and ambulance transportation reports generated by the members under his command prior to forwarding the completed reports to administration.

C. Reports shall be entered, printed, and signed prior to the end of each shift. Incident reports for EMS calls shall be entered into the computer but need not be printed. The completed reports shall be forwarded to administration.

D. The department's management information officer, EMS program manager, and fire marshal shall review all reports prior to filing and submittal to the state fire marshal to ensure accuracy and compliance with state and local guidelines. A list of errors and omissions shall be forwarded to the fire chief each month for his evaluation.

E. The fire chief shall be responsible for causing the creation of a monthly report that provides an analysis of the incident activity for the month. This report shall include cumulative data for the calendar year.

General Administration, SOP 203.05

PATIENT TREATMENT FORMS

I. **Scope**

This standard establishes the requirement that a written report be completed for every person evaluated, treated, or transported to a health care facility for medical purposes by any member of the department. It was promulgated to:

A. Establish the requirement that an accurate, complete written report be produced for all medical and rescue incidents and that the report be entered into the department's permanent record file.

B. Provide an accurate and concise evaluation of the physical and mental condition of each patient treated by the department, as well as an account of the treatment rendered to all persons during the course of an incident.

C. Provide a database for an analysis of the community's demand for emergency medical and rescue service.

D. Provide information for patient billing purposes and to facilitate the collection of a fee for services rendered by the department.

II. Completing the Reports

A. A Basic EMS Report shall be completed each time an ambulance is dispatched to an incident involving a person in need of medical evaluation, treatment, or transportation. A separate report shall be completed for each patient.

B. If possible, reports shall be filed electronically using the software program that has been installed on the department's computer system for this purpose. If this is not possible, a report may be filed manually by completing a hard copy of the required form.

C. It is unnecessary to complete a report for a fire incident if an ambulance responded as part of the initial alarm assignment and there were no injuries or fatalities.

D. An EMS Charge Sheet shall also be completed for each patient whenever treatment involves chargeable items and/or procedures. (This will vary by jurisdiction, since many departments do not charge for services.) The EMS Charge Sheet shall be securely attached to the patient's Basic EMS Report.

III. Responsibilities

A. The highest-ranking member assigned to an ambulance crew shall be responsible for the proper completion of all forms and reports. The senior member of the crew shall complete the reports if all members hold the same rank.

B. All reports and forms shall be completed prior to the conclusion of a member's tour of duty. The reports shall be accurate and thorough, and they shall contain sufficient information to allow the reader to re-create an accurate portrayal of the facts and events surrounding an incident.

C. The watch commander shall be responsible for reviewing the reports and forms completed by medical personnel under his command for accuracy and completeness prior to forwarding the reports to administration.

D. The management information system officer and EMS program manager shall review all reports to ensure accuracy and compliance with departmental policies and procedures. A list of errors and omissions shall be forwarded to the fire chief each month for his evaluation and review.

E. The fire chief shall be responsible for causing the creation of a monthly EMS report that provides an analysis of the incident activity for the month. This report shall include cumulative data for the calendar year.

Section 300

Hazardous Materials

Hazardous Materials, SOP 300.01

PROGRAM MANAGEMENT

I. **Scope**

This standard defines the parameters of the department's efforts to manage the hazardous materials problem within the community, both before and after the incident. It was promulgated to:

A. Establish a program to identify occupancies that store, use, manufacture, or distribute hazardous materials in an effort to prevent incidents from occurring that could potentially involve a hazardous material; and

B. Establish guidelines to be followed by the department for managing an incident that involves a hazardous material.

II. **General**

A. The potential always exists within the community for an incident to occur that involves one or more hazardous materials. The frequency of occurrence and the fiscal resources of the district do not, however, allow the department to operate beyond the first-responder operational level as defined by NFPA 472, *Professional Competence of Responders to Hazardous Materials Incidents.*

B. At the operational level, members are responsible for protecting people, property, and the environment from the effects of a hazardous material. Members will operate in a defensive role to contain the incident and prevent a release of a hazardous material from spreading.

C. After the incident commander has secured the scene, a private vendor shall be called to the incident to control the incident and clean up the site.

D. The private vendor may be furnished by the party responsible for the incident; otherwise, the department will contact a contractor from the approved vendor list.

1. The responsible party shall be billed for all expenses incurred in resolving the incident.
2. The local ordinance granting this authority:

____

III. **Responsibilities**

A. The emergency management coordinator shall be responsible for the management and oversight of the department's hazardous materials program and shall be responsible for ensuring that the department is in strict compliance with all applicable state and federal regulations.

B. The fire marshal shall be responsible for:
   1. Identifying businesses and other concerns that are involved in the manufacture, use, storage, or distribution of hazardous materials.
   2. Causing the periodic inspection of these businesses so as to reduce the possibility that an incident might occur.
   3. Providing a list of these businesses to the company officers for use in preparing tactical surveys of these occupancies.

C. The hazardous materials officer shall equip each apparatus with a pair of binoculars, the appropriate manuals and reference materials, portable monitoring devices, and a supply of disposable chemical protective suits.

D. The shift training officer shall be responsible for training the members assigned to his respective shift to the first-responder operational level as outlined in NFPA 472.

E. Company officers shall be responsible for:
   1. Preparing tactical surveys for occupancies within their first-due areas that manufacture, store, use, or distribute hazardous materials. During the preparation of the tactical survey, the company officer shall secure permission from the business owner or manager to affix an NFPA 704, *Standard System for the Identification of the Hazards of Materials for Emergency Response* placard in a conspicuous place to identify the most severe hazard present in the occupancy.
   2. Ensuring that their driver inspects the manuals, equipment, and protective clothing assigned to their apparatus for use in a hazardous materials incident at the beginning of each shift.
   3. Ensuring that all members under their command maintain their training and skill levels at the first-responder operational level as defined by NFPA 472.

Hazardous Materials, SOP 301.01

EMERGENCY RESPONSE

I. Scope

This standard establishes guidelines to be followed during the management of an incident involving a hazardous material. It was promulgated to:

A. Establish guidelines for the management of a hazardous materials incident.

B. Provide for the safety of response personnel.

II. Definitions

The following definitions are taken from a variety of sources including NFPA 472, *Standard for Professional Competence of Responders to Hazardous Materials Incidents.*

A. Cold zone: The zone of a hazardous materials incident that contains the command post and such other support functions as are deemed necessary to control the incident. This zone is also referred to as the clean zone or support zone.

B. Confinement: Those procedures taken to keep a material in a defined or local area once released.

C. Containment: The actions taken to keep a material in its container (e.g., stop the release of the material or reduce the amount being released).

D. Contaminant: A hazardous material that physically remains on or in people, animals, the environment, or equipment, thereby creating a continuing risk of direct injury or a risk of exposure outside of the hot zone.

E. Decontamination: The physical or chemical process of reducing and preventing the spread of contamination from persons and equipment used at a hazardous materials incident.

F. Exposure: The process by which people, animals, the environment, and equipment are subjected to or come in contact with a hazardous material.

G. Hazardous material: A substance that when released is capable of creating harm to people, the environment, animals, and property.

H. Hot zone: The area immediately surrounding a hazardous materials incident. It extends far enough to where adverse effects from hazardous materials release will not be expected. It is also referred to as the exclusion zone or the restricted zone.

I. Penetration: The movement of a material through a suit's closures—such as zippers, buttonholes, seams, flaps, or other design features of chemical-protective clothing—and through punctures, cuts, and tears.

J. Stabilization: The point in an incident at which the adverse behavior of the hazardous material is controlled.

K. Warm zone: The control zone at a hazardous materials incident where personnel and equipment decontamination and hot zone support takes place. It includes control points for the access corridor, helping to reduce the spread of contamination. This zone is also referred to as the decontamination, contamination reduction, or limited access zone.

III. General

A. The first-arriving officer at an incident involving hazardous materials shall report the following information to Dispatch:

1. The exact location of the incident.
2. The type and quantity of the materials involved, if known.
3. The extent of damage and the number and types of injuries.
4. The name of the carrier if a vehicle is involved.
5. Any other pertinent information such as the hazardous material entering the storm drain or sanitary sewer system.

B. The first-arriving officer shall also request any additional resources that may be needed, establish command, and begin securing the incident scene to prevent additional injuries or contamination.

C. The area shall be evacuated if necessary. This function should be turned over to the police once sufficient law enforcement resources arrive on the scene.

D. Dispatch shall provide the incident commander with the temperature, wind speed and direction, and humidity as soon as it is possible to do so.

IV. Tactical Objectives

A. The incident commander shall assess the situation and identify the product(s) involved prior to committing personnel.

B. If entry must be made into the hot zone to rescue someone or to contain a release, the appropriate level of protective clothing shall be worn by the personnel who enter.

C. If personnel are committed to the hot zone, decontamination and rehab sectors shall be established, as well as a warm zone and a cold zone.

D. Decontamination of victims shall also occur prior to their being transported to a medical facility.

E. Additional resources shall be requested in accordance with the district's Emergency Operations Plan. This includes the dispatch of a private vendor if the incident exceeds the department's capabilities.

F. The incident commander's objectives will be containment and stabilization. Final extinguishment and cleanup of incidents that exceed the capabilities of the department will be the responsibility of the private vendor who responds to the incident.

G. In more complicated incidents, it may be necessary to activate the Emergency Operations Plan.

Hazardous Materials, SOP 301.02

CONTAINMENT AND CLEANUP

I. Scope

This standard establishes guidelines for managing the containment and cleanup of a hazardous materials incident.

II. General

A. Containment methods may include but shall not be limited to:

1. Barriers in soil.
2. Berms and drains.
3. Booms.
4. Dikes.
5. Diverting streams.
6. Overpacked drums or other forms of containerization.
7. Patching and plugging of containers or vessels.
8. Portable catch basins.
9. Reorienting the container.
10. Trenches.

B. Displacement techniques may include but shall not be limited to:

1. Dispersion/dilution.
2. Excavating.
3. Hydraulic and mechanical dredging.
4. Skimming.
5. Vacuuming.

III. Responsibilities

A. The incident commander, in coordination with the appropriate state or federal official, is responsible for selecting and implementing the appropriate countermeasures to bring a hazardous materials incident to a safe and successful conclusion. This includes:

1. Ensuring that temporary storage sites, if necessary, are safe and secure.
2. Ensuring that final disposal is handled at an approved site.

B. (Insert local regulations about contacting the city, county, or district health department and the appropriate state and federal agencies—e.g., the EPA. Describe also the scenarios under which each agency should be summoned.)

C. The persons responsible for the hazardous material are responsible for paying the full costs for cleanup and disposal operations.

Hazardous Materials, SOP 301.03

DECONTAMINATION PROCEDURES

I. Scope

This standard establishes a procedure for the decontamination of people, equipment, and apparatus that become contaminated as a result of an exposure to a hazardous material. It was promulgated to:

A. Prevent the spread of contaminants beyond the hazard zone at an incident involving a hazardous material.

B. Reduce the possibility of death or injury due to exposure to a hazardous material.

C. Establish a procedure to decontaminate equipment and apparatus exposed to a hazardous material so that the equipment and apparatus might promptly be returned to service.

II. General

A. Contamination is the transfer of a hazardous material to persons, equipment, and the environment due to an exposure or contact with a hazardous material. The magnitude of the exposure depends on the duration of the exposure and the concentration of the hazardous material.

B. Decontamination (decon) is the process of removing contaminants from people and equipment. Decon should occur in the warm zone so as to minimize the possibility of secondary contamination.

C. There are four basic methods of decontamination available to response personnel:

1. Dilution: The use of water to flush the contaminant from the victim or piece of equipment. Be sure to impound or collect the contaminated water from this process.
2. Absorption: The use of an absorbent for picking up a liquid contaminant. This works well on a spill.
3. Chemical degradation: The use of another material (e.g., household bleach or baking soda) to change the chemical structure of the hazardous material so as to neutralize the material.
4. Isolation and disposal: The isolation of a hazardous material by collecting it and then disposing of it in accordance with state and federal regulations. While this may be a more costly alternative, it is often the easiest technique to employ.

III. Procedure

A. If contamination has occurred at a hazardous materials incident, the incident commander shall appoint a decon officer. The decon officer shall establish a decontamination sector in the warm zone. The site of the decon sector should be selected on the basis of:

1. Accessibility and location. (Special note: Due consideration must be given to the privacy of potential victims and rescue workers.)
2. Surface material.
3. Lighting.
4. Drains and waterways.
5. Water supply.
6. Weather.

B. All personnel, victims, and equipment must be decontaminated prior to being allowed entry into the cold zone. The specific decontamination measures employed in the decon sector will depend on the circumstances surrounding the incident and the level of contamination.

C. Members assigned to the decon sector will instruct contaminated members to follow the following procedure:

1. Gross decontamination: A decontamination worker will remove the majority of the contamination from the victim and his tools by hosing down the victim or by providing a portable shower. A catch basin shall be used to confine the water used in this process.
2. Tools and equipment: Any tools and equipment used by the victim should then be discarded at a designated location for further decontamination, if necessary.
3. Scrubdown: The victim should then step into the rinse area, where a decontamination worker will scrub him with detergent and water. The water shall be kept for analysis prior to being released.
4. Final step: After being washed down, the victim should then proceed to the final area, where articles of clothing and other equipment will be removed. These items shall be left in the decontamination area for further treatment or disposal.

D. If a contaminated victim must be transported prior to being properly decontaminated, Medical Control and the destination emergency room shall be notified.

E. If an ambulance becomes contaminated, the ambulance will be quarantined until the unit can be properly decontaminated.

F. Members who are exposed to a hazardous material shall complete a Hazardous Materials Exposure Form and place a copy of it in their medical file.

Hazardous Materials, SOP 301.04

PERSONNEL SAFETY

I. Scope

This standard establishes guidelines for members to follow when engaged in an incident involving a hazardous material. It was promulgated so as to minimize the risk of death and injury during haz-mat incidents.

II. General

A. The most important action to be taken at a hazardous materials incident is to recognize that a hazardous material is present. If the potential exists, assume the worst until it can be confirmed that no danger exists.

B. The area must be secured and no one must be allowed to enter the area until the incident commander determines that it is safe to do so.

C. Appropriate help should be summoned and the material or materials present should be identified.

D. Once the problem has been identified, the incident commander should formulate an action plan to resolve the incident. Tactical surveys will be useful in this process. The action plan should focus on three goals:

1. Life safety.
2. Environmental protection.
3. Property conservation.

E. The incident commander should appoint a safety officer early in this process. The safety officer must remain in constant contact with the IC and should immediately correct any unsafe conditions or practices.

III. Operations

A. There are three recognized levels of response:

1. Level I: The least serious and within the capabilities of the department. Evacuation, if required, will be limited to the immediate area. Example: a gasoline or diesel spill.
2. Level II: Beyond the capabilities of the department, a Level II incident requires the service of a formal Haz Mat Response Team. Examples: the rupture of a pipeline or a fire with the threat of a BLEVE.
3. Level III: The most serious type, requiring special resources from public and private agencies. It will require a large-scale evacuation and implementation of the jurisdiction's Emergency Operations Plan. Example: an incident that extends across jurisdictional boundaries.

B. Operations will either be defensive or offensive.

1. Defensive operations focus on confinement without directly contacting the hazardous materials creating the problem.
2. Offensive operations focus on aggressive actions on the material, container, or process, and they may result in contact with the materials. Offensive operations will not be conducted unless the incident commander can determine that the risk is worth the benefit.

C. Personnel will not be allowed to enter the warm zone or the hot zone without the proper level of training and personal protection.

D. Personnel will not be permitted to leave the warm zone or the hot zone without undergoing proper decontamination.

E. The safety officer will appoint a medical surveillance team. The team will be responsible for monitoring members for indicators of toxic exposure effects, including:

1. Change in complexion, skin discoloration.
2. Lack of coordination.
3. Changes of demeanor.
4. Excessive salivation, papillary response.

5. Changes in speech pattern.
6. Headaches.
7. Dizziness.
8. Blurred vision.
9. Cramps.
10. Irritation of eyes, skin, or respiratory tract.

F. Any member exposed to a hazardous material shall complete a Hazardous Materials Exposure Form. A copy of the form shall be placed in the member's medical file.

G. The incident commander shall also appoint a rehab officer. The rehab officer shall establish a rehab sector. Members shall be rotated through the rehab sector in accordance with the department's SOPs.

Hazardous Materials, SOP 301.05

CONTAMINATED PROTECTIVE CLOTHING

I. **Scope**

This standard establishes guidelines to be followed when uniforms, protective clothing, or items of personal protective equipment are contaminated by a hazardous material.

II. **Procedure**

A. Personnel who enter the warm zone or the hot zone at a hazardous materials incident run the risk of becoming contaminated by the materials involved.

B. Members must pass through the decontamination sector prior to being allowed to enter the cold zone.

1. At the entrance to the decon sector, the member should discard any hand tools and equipment at the edge of the corridor so that they can be decontaminated.
2. After the member has been rinsed off and decontaminated as much as possible, he shall proceed to the final area where a decontamination worker will assist him in removing his protective clothing.
3. All articles of contaminated protective clothing shall be placed in a bag, then sealed and tagged. The tag should list the contaminant, the contents of the bag, the member's name, and the time and date.
4. If a member's personal clothing or work uniform has been contaminated, the member should proceed to a showering station. After showering, he should dry off his body and change into clean clothes. Ensure that the member is afforded all due privacy.
5. Contaminated uniforms and articles of personal clothing should be bagged, then sealed and tagged. Towels and other items used to dry off should also be placed in a bag for decontamination or disposal.

C. The decon officer shall make a determination if contaminated items are salvageable. Items that cannot be properly decontaminated shall be disposed of in accordance with state and federal regulations.

D. Items that may be decontaminated shall be returned to the member after being properly decontaminated.

Section 400
Occupational Safety and Health

Occupational Safety and Health, SOP 400.01

OCCUPATIONAL SAFETY AND HEALTH PROGRAM

I. Scope

This standard establishes an occupational safety and health program for the department. It was promulgated to:

A. Provide a safe working environment for the members of the department.

B. Satisfy the requirements of NFPA 1500, *Standard on Fire Department Occupational Safety and Health Program.*

II. Policy Statement

It shall be the policy of the department to operate at the highest possible level of safety and health for all its members. To this end, the department shall:

A. Make every reasonable effort to provide a safe and healthy work environment.

B. Give primary consideration to the prevention and reduction of accidents, injuries, and occupational illnesses.

C. Take the appropriate corrective action to avoid repetitive occurrences of accidents.

D. Provide training, supervision, written procedures, program support, and review for all of its activities.

III. Responsibilities

A. Safety and health is the responsibility of every member. Therefore, each member shall:

1. Cooperate, participate, and comply with all of the provisions of the occupational safety and health program.
2. Promptly report acts and conditions that are unsafe or unhealthy and that pose a threat either to members or to others.
3. Maintain a level of mental and physical fitness that enables the member to safely perform his assigned tasks.

B. Supervisors shall be responsible for enforcing the requirements of the occupational safety and health program and for ensuring that each member under their command complies with the provisions of the occupational safety and health program.

Occupational Safety and Health, SOP 400.02

FIRE DEPARTMENT SAFETY OFFICER

I. Scope

This standard establishes the position of safety officer. It was promulgated to assign the responsibility for the department's occupational safety and health program to a single individual.

II. General

A. The fire chief shall appoint a member to serve as the department's safety officer. This shall be the officer's collateral duty as specified in SOP 201.02, Collateral Assignments.

B. The department's safety officer shall:

1. Acquaint himself with the provisions of NFPA 1500, *Standard on Fire Department Occupational Safety and Health Program,* and NFPA 1521, *Standard for Fire Department Safety Officer.*
2. Manage the department's occupational safety and health program.

C. The duties of the department's safety officer shall include but not be limited to:

1. Identifying safety and health hazards and developing plans to correct them.
2. Immediately correcting situations that create an imminent hazard to the members of the department.
3. Training members to act as incident safety officers.
4. Maintaining records of accidents, occupational deaths, injuries, illnesses, and exposures and providing analysis and reports to the fire chief as directed.
5. Acting as the department's liaison with the district's risk manager, as well as the district's safety committee.
6. Serving as the chair of the department's safety committee. In this capacity, the safety officer shall prepare and distribute meeting agendas and notices and shall forward a copy of the minutes of each meeting to the fire chief.
7. Serving as a member of the department's Driver Accident Review Board.
8. Providing safety training, bulletins, posters, and newsletters to all members.
9. Performing other duties as specified in NFPA 1521 or as directed by the fire chief.

Occupational Safety and Health, SOP 400.03

OCCUPATIONAL SAFETY AND HEALTH COMMITTEE

I. Scope

This standard establishes the department's Occupational Safety and Health Committee. It was promulgated to:

A. Establish an advisory committee for occupational safety and health.

B. Establish guidelines for committee membership.

C. Assign duties and responsibilities to committee members.

II. General

A. The Occupational Safety and Health Committee shall be responsible for conducting research, developing recommendations, and studying and reviewing matters pertaining to occupational safety and health.

B. The committee shall be composed of a minimum of four members, not including the chair. Members shall be appointed in such a manner that there is at least one representative from each suppression shift and one representative from administration/prevention.

C. The suppression representatives shall include at least one firefighter, one driver, and one officer. At least one of the members must be a paramedic.

III. Responsibilities

A. The department's safety officer shall serve as the chair of the committee and shall, with the approval of the fire chief, appoint the other members of the committee.

B. The chair of the Occupational Safety and Health Committee shall schedule and hold regular meetings of the committee. Meetings shall be conducted at least quarterly. Special meetings may also be conducted as required.

C. The chair shall ensure that written minutes of each meeting are kept. Copies shall be made available to the chief and the committee members. A copy of the minutes shall also be posted on the bulletin board of each fire station.

D. Committee members shall attend scheduled meetings and assist the safety officer in the operation of the department's safety program.

E. The committee shall review all accident and injury reports and make recommendations concerning its findings.

F. The committee shall maintain a working knowledge of applicable safety rules and regulations and shall recommend programs, practices, methodologies, and changes to reduce or eliminate accidents and injuries.

Occupational Safety and Health, SOP 400.04

RECORDS

I. Scope

This standard establishes a system for the collection and permanent retention of information concerning accidents, injuries, illnesses, exposures to infectious agents and communicable diseases, and deaths that are or might be job related. It was promulgated to:

A. Establish the requirement that a report be completed and filed for every accident, injury, illness, exposure to an infectious agent or communicable disease, or death that involves a member while on duty or that may in some way be related to the member's job.

B. Establish a database for analysis to develop programs to reduce on-the-job accidents, illnesses, and deaths.

II. General

A. A permanent record of all accidents, injuries, illnesses, or deaths that occur while a member is on duty shall be maintained by the department. The record shall include both a master file of such events as well as individual records placed in the individual member's permanent personnel file.

B. Other records and reports that are useful in evaluating the department's overall safety and health program shall be maintained as required by rule, regulation, policy, or standard operating procedure.

C. A tabulation of the department's safety record shall be included in the monthly report and the annual report.

III. **Responsibilities**

A. It shall be the responsibility of each member of the department to immediately report to his supervisor any accident, injury, or illness that occurs while on duty. The appropriate report shall be filed in writing within 24 hours of the event.

1. If a member was exposed to a hazardous material or contagious disease, he shall file a Hazardous Materials Exposure Form as required by SOP 406.03, Exposure Reporting.
2. A district Notification of Accident Report shall be filed if the member was involved in a motor vehicle accident.
3. If a member has been injured, an On-the-Job Injury Supervisor's Investigation Report and a Supplemental Report of Injury (Workers' Comp) report shall be filed.
4. If the member is treated by a physician, a Designation of Duty Status Form shall also be completed.

B. Officers shall be responsible for ensuring that a member under their command who is involved in an accident or injured completes the appropriate reports. Reports shall be complete, accurate, and filed in a timely manner.

C. The department's safety officer shall review all accident and injury reports and shall ensure that the event is reviewed by the appropriate committee for its recommendation.

Occupational Safety and Health, SOP 401.01

OPERATING HYDRAULIC-POWERED RESCUE TOOLS

I. **Scope**

This standard mandates safety guidelines to be followed while operating a hydraulic-powered rescue tool. It was promulgated to:

A. Prevent accidents, injuries, and deaths that might result from the misuse or improper operation of hydraulic-powered rescue tools.

B. Prevent damage to hydraulic-powered rescue tools that might result from misuse or abuse.

II. **General**

A. Members that respond to fire and emergency medical incidents are responsible for knowing how to properly and safely operate hydraulic-powered rescue tools.

B. Officers shall train the members under their command to properly and safely operate the hydraulic-powered rescue tools assigned to their apparatus.

C. An officer shall immediately stop any unsafe or improper operation of a hydraulic-powered rescue tool and make the adjustments and corrections necessary to safely accomplish the assignment.

D. Drivers shall inspect the hydraulic-powered rescue tools assigned to their apparatus at the beginning of each tour of duty and after every use to ensure that they are functioning properly. Tools found to be unsafe or malfunctioning shall be removed from service and be properly red-tagged.

E. Drivers shall check the fuel level in each tool to ensure that each is properly fueled. The spare fuel can carried on the apparatus shall also be kept full at all times.

III. Operating Procedure

A. Before operating a tool, always inspect the tips to make sure that the appropriate tips are being used and that the retainer pins, if used, are in place.

B. Place the power unit as level as possible, connect the hoses, but do not start the power unit until the tool operator gives the command to do so.

C. Remember that it takes two people to operate the tool: one person to operate the tool and another person to operate the power unit.

D. To start the power unit, place one foot on the bottom of the roll cage to help stabilize the unit. Hold the cage with one hand and pull the recoil starter cord with the other. Take care not to pull the cord out too far. The start-up sequence is as follows:

1. Move the choke to the closed position.
2. Pull the recoil starter until the engine pops.
3. Move the choke to run.
4. Pull the recoil starter cord.
5. Repeat the sequence if the engine fails to start.

E. To stop the power unit, use your hand to engage the kill switch. Do *not* use your foot.

F. When refueling the tool, always use the appropriate fuel type and take care not to spill fuel on a hot surface. Note: The pitch of the power unit will normally change prior to running out of fuel.

G. Safety precautions:

1. Full protective clothing, including ear protection, shall be worn while using a hydraulic-powered rescue tool during actual rescues as well as training exercises.
2. Ear protection, approved safety shoes or boots, and any other item of protective clothing that is appropriate shall be worn while operating a tool for the purpose of inspection or maintenance.
3. Always work on the outside of the tool. *Never* insert your hands or other parts of your body between the jaws or tips of the tool.
4. Cover and protect the victim, and always explain to him what you're doing, if possible.
5. When bleeding or disconnecting a hose, place a rag over the coupling to prevent fluid from spraying on anyone.
6. If fluid comes in contact with any exposed skin area, be sure to wash it off immediately.
7. If a fluid spill does occur, immediately clean any floor area or painted surface to prevent a fall or damage.
8. Always remember to make the required hose connections prior to starting a power unit. Hose should be laid out in such a manner as to prevent damage from sharp objects, vehicles, etc.

9. Likewise, always stop a power unit prior to disconnecting a hose.
10. Do not use the shears to cut a steering column or any piece of metal with a free end, since it may become a projectile. Shears are designed to create a compression fracture rather than to cut. Always remember to cut at a right angle.
11. At a motor vehicle accident, always have a charged hoseline to protect against the possibility of a fire.
12. Rotate personnel to avoid fatigue.
13. Rest the tool on your thigh and always maintain body balance.
14. Guide and hold the tool; do not force it. Don't strain against the tool; rather, work with it.
15. Always respect the tool. It is a machine and has no conscience.

Occupational Safety and Health, SOP 401.02

OPERATING POWER SAWS

I. Scope

This standard establishes guidelines for the safe and proper operation of power saws. It was promulgated to:

A. Prevent accidents, injuries, or deaths that might result from the improper use or unsafe operation of a power saw.
B. Prevent damage to a power saw that might result from unsafe operation or improper use.

II. General

A. Members who respond to fire and emergency medical incidents shall be responsible for knowing how to properly and safely operate the power saws used by the department.
B. Officers shall train the members under their command in the safe and proper use of the power saws assigned to their apparatus.
C. An officer shall immediately stop any unsafe or improper use of a power saw and shall take the appropriate action necessary to correct the situation.
D. Drivers shall inspect each power saw carried on their apparatus at the beginning of their tour of duty and after each use to ensure that they are clean, functioning properly, and safe to operate. Any power saw discovered to be unsafe or malfunctioning shall be removed from service and properly red-tagged.
E. Drivers shall also be responsible for ensuring that saws are properly fueled and that spare fuel containers are full of the correct gas/oil mixture and in the proper place on the apparatus.

III. Operating Procedures

A. Always carry a power saw with the engine stopped or the electrical power disconnected. The blade should be carried to the front with the muffler away from your body.
B. Always keep both hands on the control handles, using a firm grip with your thumbs and your fingers encircling the handles.
C. Make sure of your footing prior to operating a saw.
D. Always turn off a saw when it is unattended.

E. Have a plan of action before placing a saw into operation. The plan should include:
   1. The location and sequence of the cuts and openings.
   2. Wind direction—consider its effects on exposure and personnel.
   3. Escape routes—at least two means of egress.

F. Always place the safety guard in the proper position for the use intended before operating the saw.

G. Remember that power saw operations are safest when cutting on a horizontal surface near the ground level or on a vertical surface at or below waist level.

H. Operating a power saw above chest height is extremely hazardous and should not be attempted as a normal course of action. This type of operation should be conducted only under the direct order and supervision of an officer. The officer ordering this operation shall first consider the value to be gained vs. the extreme hazard to personnel.

I. The use of a power saw from a ladder shall only be done if no other alternatives are available.

J. Do not operate a power saw close to a highly combustible or flammable material due to the possibility of ignition from sparks.

K. Do not operate saws in flammable or explosive atmospheres.

L. When operating a power saw, avoid placing side pressure or twisting the blade. Never force the saw. If too much pressure is applied to the blade, the hazard of blade breakage (carbide tipped) or shattering (aluminum oxide or silicon carbide discs) is increased. A blade that breaks or shatters during cutting operations may cause serious injury to the operator or bystanders.

M. The saw cut should only be as deep as necessary. Deep cuts may weaken supporting beams and lead to collapse. The experienced operator will know when he has reached a beam by the sound and feel of the saw.

N. If conditions permit, scrape gravel and debris from the cutting path to reduce the danger of injury from flying chips and loose materials.

IV. **Safety Precautions**

A. A member who operates a power saw at an emergency incident or during a training session shall wear full protective clothing, including both ear and eye protection.

B. A member who operates a power saw for the purpose of inspection or maintenance shall wear ear protection and safety shoes or boots.

C. When operating a power saw, all clothing shall be close fitting and completely buttoned to prevent an accident due to moving belts, gears, chains, blades, etc.

D. Do not operate a gasoline-powered saw with a fuel leak. Remove the saw from service.

E. Do not restart a saw in a small enclosed space after refueling.

V. **Fueling and Maintenance**

A. Power saws shall be kept clean and in good serviceable condition.

B. The cutting wheel, chain, or blade shall be examined at the beginning of each tour of duty and after each use for nicks or defects. These items should also be checked for tightness and shall be kept clean and properly lubricated. Defective items shall be replaced.

C. Ensure that abrasive saw blades do not become contaminated with petroleum-based products. Such contamination may dissolve the resin that is used to bond the blade, thus causing the blade to shatter when used. New blades should be stored in plastic bags to ensure cleanliness.

D. When fueling a power saw:

1. Always turn the engine off.
2. Make sure to use the proper fuel mixture. Many saws require a specific fuel and oil mixture.
3. Wipe off the saw to remove any spilled fuel before starting it.

Occupational Safety and Health, SOP 402.01

TRAINING

I. Scope

This standard applies to all training conducted by or for the department. It was promulgated to:

A. Prevent occupational accidents, deaths, injuries, and illnesses.

B. Ensure that all members are able to properly perform their assigned duties in a safe manner.

II. General

A. The department shall provide training to its members to update them on new practices and techniques and to help them maintain individual skill levels.

1. Sufficient training will be scheduled each calendar year to allow members to maintain their EMT or paramedic certification.
2. Sufficient training shall also be scheduled each calendar year to allow members to maintain their firefighter certifications.

B. All emergency medical training shall be approved by the emergency medical services program manager and shall be taught by an instructor determined by the program manager to be qualified.

C. All fire suppression and rescue training shall be approved by the fire chief and shall be conducted under the supervision of a certified instructor.

D. Each shift has a designated shift training officer who has been assigned to supervise training on his respective shift and to maintain records of the training received by each member. See SOP 202.01, Shift Training Officers.

E. All training involving live-fire exercises shall be conducted in compliance with the provisions of NFPA 1403, *Standard on Live Fire Training Evolutions.*

III. Responsibilities

A. Each member shall be responsible for maintaining his fire and EMS skills at a level sufficient to retain the certifications required for his job and to meet the established minimum standards of performance. See SOP 600.03, Minimum Company Standards.

B. Officers shall ensure that all members under their command maintain their certification and skill levels. The officer shall also be responsible for maintaining his company's certification and skill at a level sufficient to meet the established minimum standards of performance. See SOP 600.03, Minimum Company Standards.

Occupational Safety and Health, SOP 402.02

SCBA TRAINING

I. <u>Scope</u>

This standard applies to all members who are required to use self-contained breathing apparatus (SCBA) as a part of their normal duties. It was promulgated to:

A. Maintain individual proficiency in the use of SCBA.

B. Prevent accidents, injuries, and deaths that might result from exposure to a hazardous atmosphere.

II. <u>Training and Evaluation</u>

A. Each member of the department required to use SCBA as a part of his normal duties shall receive periodic training on the proper use of SCBA. Training shall be based on the requirements of NFPA 1404, *Standard on Fire Department Self-Contained Breathing Apparatus Program.*

B. Each member shall be evaluated on an annual basis to ensure that he is proficient in the use of SCBA. As a part of their evaluation, members shall successfully:

1. Identify the components of facepieces, regulators, harnesses, and cylinders.
2. Correctly don, operate, and doff SCBA while wearing full protective clothing. This shall include demonstrating that a proper facepiece seal has been achieved.
3. Describe the operational principles of the warning devices.
4. Identify the limitations of SCBA, correctly define the term "point of no return," and discuss the ability to protect the body from absorption of toxins through the skin.
5. Describe the procedures to be used if unintentionally submerged in water while wearing SCBA.
6. Demonstrate alternative means of communication while wearing SCBA.
7. Demonstrate the procedure for daily inspection and maintenance of SCBA.
8. Demonstrate the procedure for cleaning and sanitizing SCBA for future use.
9. With SCBA donned, perform related emergency scene activities such as advancing a hoseline, climbing a ladder, crawling through a window or confined space, performing a rescue, etc.
10. Conduct an annual facepiece fit test.

III. <u>Responsibilities</u>

A. Each member of the department required to use SCBA as a part of his normal duties shall strictly adhere to the requirements of this standard.

B. Company officers shall ensure that each member assigned to their command strictly adheres to the requirements of this standard and shall conduct the training and evaluation required by this standard.

C. The shift training officers shall maintain records of any SCBA training provided to the personnel assigned to their respective shifts. At least once each year, shift training officers shall evaluate the ability of each member to meet the requirements imposed by this standard. The evaluation shall be conducted as a part of the annual Minimum Company Standard testing. See SOP 600.03, Minimum Company Standards.

Occupational Safety and Health, SOP 402.03

SABA TRAINING

I. **Scope**

This standard applies to all members who may be required to use a supplied air breathing apparatus (SABA) as part of their normal duties. It was promulgated to:

A. Maintain individual proficiency in the use of SABA.

B. Prevent accidents, injuries, and deaths that might result from exposure to a hazardous atmosphere.

II. **Training and Evaluation**

A. Each member of the department required to use a SABA as part of his normal duties shall receive periodic training on the proper use of SABA.

B. Each member shall be evaluated on a semiannual basis to ensure that he is proficient in the use of SABA. As a part of their evaluation, members shall successfully:

1. Identify the major components of the air-source cart and the air line/egress respirator.
2. Demonstrate their ability to place the cart in service.
3. Change cylinders while the cart is in use by using the air bleed valve.
4. Activate and deactivate the electronic command module warning system.
5. Connect and disconnect the air line at the low-pressure manifold by using the connection's safety device.
6. Don the air line/egress respirator and connect it to the supplied air line.
7. Place the egress respirator in service when the supplied air is lost.
8. Shut down all components of the cart, including the electronic command module and all of the warning systems.
9. Demonstrate the procedures for care and maintenance of the cart and the air line/egress respirator.

III. **Responsibilities**

A. Each member of the department required to use SABA as part of his normal duties shall strictly adhere to the requirements of this standard.

B. Company officers shall ensure that each member assigned to their command strictly adheres to the requirements of this standard and shall conduct the training and evaluation required by this standard.

C. The shift training officers shall maintain records of any SABA training provided to the members assigned to their shift.

D. In addition, the shift training officers shall evaluate the ability of each member assigned to their shift to meet the requirements imposed by this standard semiannually.

Occupational Safety and Health, SOP 403.01

DRIVERS OF VEHICLES

I. Scope

This standard applies to all members who drive or operate a motorized vehicle owned or used by the department. It was promulgated to:

A. Establish minimum standards for members who are allowed to drive or operate a motorized vehicle.

B. Establish minimum safety regulations for the operation of a motorized vehicle.

II. General

A. Driver's license:

1. All members of the department shall have a valid driver's license that is appropriate for the types of vehicles that they are allowed to operate.
2. All members shall furnish proof of their possession of a valid license anytime that they are requested by a supervisor to do so.
3. Supervisors shall check the driver's license of each member under their command during the first shift in January and in July to determine that each member possesses a valid license.
4. Members shall report any change in the status of their driver's license to their supervisor. The supervisor shall forward this information to Administration via the normal chain of command.
5. Any member who has his license suspended shall notify his supervisor immediately and shall not be allowed to drive or operate a vehicle until his license has been restored. Failure to possess a valid driver's license shall be grounds for suspension.

B. For the purposes of this standard, motorized vehicles shall be divided into two categories: Category One and Category Two.

1. Category One vehicles are those vehicles used primarily for fire suppression purposes and that have a gross vehicle weight (gvw) that exceeds 11,000 lbs.
2. Category Two vehicles are those vehicles that have a gvw of 11,000 lbs. or less and are primarily used for EMS or support services.

C. No member shall be allowed to drive or operate a Category One vehicle unless he has successfully completed the department's driver training program or is a student driver under the supervision of a qualified driver.

D. All vehicles shall be operated in a safe and prudent manner, and all drivers shall comply with all traffic laws and the applicable rules and regulations of the department.

E. No driver shall move a vehicle until all persons in it are in an approved riding position and are properly secured.

F. Drivers responding to emergencies shall comply with the provisions of SOP 403.03, Emergency Response.

G. A driver shall *not* back a vehicle unless his view is clear and unobstructed. Apparatus and ambulances shall not be backed unless there is at least one spotter to the rear of the apparatus to assist in the operation.

III. Responsibilities

A. Drivers shall be directly responsible for the safe and prudent operation of their vehicles in all situations.

B. When a driver is under the direct supervision of an officer, the officer shall be responsible for the actions of the driver.

C. The fire department safety officer shall monitor the status of all the members of the department to ensure that all those who drive and operate vehicles have had the proper training, possess valid driver's licenses, are insurable, and have had a defensive driving course.

D. Drivers shall be responsible for ensuring that all of their vehicles' safety equipment is functioning properly and that their vehicles are safe to drive prior to operating them.

IV. Accidents

A. The driver of a motor vehicle shall immediately notify his supervisor and the appropriate law enforcement agency if he is involved in an accident. All reports and information concerning the accident shall be forwarded to the Vehicle Accident Review Board for review.

B. Drivers and supervisors shall be familiar with the provisions of the district's substance abuse policy. Any driver involved in an accident that falls within the scope of the policy shall be tested whenever required.

C. The supervisor shall notify the fire chief whenever an accident involves an injury, fatality, or major damage to a vehicle.

Occupational Safety and Health, SOP 403.02

PERSONS RIDING IN MOTORIZED VEHICLES

I. Scope

This standard shall apply to all persons riding in or on a motorized vehicle owned or operated by the department except for EMT and paramedic students approved to ride on an ambulance. Student riders are covered by a separate standard.

II. General

A. No one shall be allowed to ride in an apparatus or ambulance unless he is a member of the department or has obtained special permission to ride as an observer. Observers must complete the Authorization to Ride an Apparatus Release.

B. Persons riding in or on a motorized vehicle shall observe the following:

1. All persons shall be seated in an approved riding position and shall be secured by a seat belt anytime the vehicle is in motion. Exception: EMT/paramedics attending a patient in the patient compartment of an ambulance shall be secured to whatever extent is practical while still being able to deliver proper medical care.
2. Riding on the tailboard, running board, or other exposed position is strictly prohibited.
3. Standing while riding is prohibited. Any member who rides in a standing position shall immediately be suspended pending an investigation of the incident.

4. At no time shall anyone dismount a vehicle while it is still in motion.
5. All persons riding on an apparatus that requires ear protection shall wear the ear protection devices provided while the vehicle is in motion and whenever the audio warning devices are in use.

III. <u>Hose-Loading Operations</u>

A. Hose-loading operations may be permitted on moving apparatus provided that the following conditions are met:
   1. A member other than those loading the hose shall be appointed as a safety observer. The safety observer shall have an unobstructed view of the hose-loading operation and shall be in both visual and voice contact with the apparatus operator.
   2. Vehicular traffic shall be excluded from the area or shall be under the control of an authorized traffic control person.
   3. Apparatus speed shall not exceed five mph when loading hose.
   4. The apparatus shall not be moved until all members involved in loading the hose have been made aware that the apparatus is about to be moved.
   5. Members in the hosebed shall not stand while the vehicle is being moved, and those members on the tailboard shall step off the apparatus prior to its being moved.

B. The safety observer shall have the authority to discontinue any hose-loading operation that is deemed to be unsafe.

IV. <u>Responsibilities</u>

A. It shall be the responsibility of each member to comply with the provisions of this standard.

B. Drivers shall not operate their vehicles unless everyone on board is in compliance with the provisions of this standard.

C. Officers shall be strictly accountable for enforcing the provisions of this standard and shall correct any violations that are observed.

Occupational Safety and Health, SOP 403.03

EMERGENCY RESPONSE

I. <u>Scope</u>

This standard applies to the driver of an emergency vehicle owned or operated by the department while responding to an incident. It was promulgated to establish safety guidelines during emergency responses.

II. <u>Categories of Response</u>

(The following are recommendations. Actual practice may vary by jurisdiction.)

A. <u>Emergency</u>: Those incidents that pose a significant risk to life or property. Emergency response requires the use of all audio (siren and airhorns) and visual (lights) warning devices. These devices must be in use during the entire duration of the response unless the response is downgraded to a nonemergency by a competent authority. The initial response to the following types of incidents shall be considered emergencies:
   1. A reported fire in a structure.
   2. A reported fire outside of a structure that involves the potential destruction of property or poses a risk to human or animal life.

3. All categories of emergency medical incidents except nonlife-threatening transfers to or from a medical facility.
4. Responses to a man-made or natural disaster involving the destruction of property and the potential for injury or death. This would include requests for assistance from other jurisdictions.

B. Nonemergency: Those incidents that do not pose a significant risk to life or property. Audio and visual warning devices shall not be used during nonemergency responses unless ordered by a competent authority to upgrade the response to emergency status. The initial response to the following types of incidents shall not be considered to be emergencies:
   1. Medical incidents that involve transfers to or from a medical facility where the patient does not have a life-threatening condition.
   2. Automatic fire alarms until confirmation is received that an actual emergency exists.
   3. Public service calls to assist the public when there is no immediate threat to life or property.

III. Response Guidelines

A. Apparatus and vehicles engaged in a nonemergency response shall obey all applicable traffic safety rules and regulations and shall not exceed the posted speed limit.

B. Apparatus and vehicles engaged in an emergency response shall at all times govern their response by the traffic, weather, and road conditions present at the time of response.

C. The maximum speed of travel shall *not* exceed posted limits by more than 10 mph.

D. During an emergency response, drivers shall bring their vehicles to a complete stop for any of the following:
   1. When directed by a law enforcement officer.
   2. Stop signs.
   3. Red traffic signals.
   4. Negative right-of-way intersections.
   5. Blind intersections.
   6. When the driver cannot account for all lanes of traffic in an intersection.
   7. When other intersection hazards are present.
   8. When encountering a stopped school bus with flashing warning lights.

E. Drivers shall proceed through an intersection only when the driver can account for all lanes of traffic in the intersection.

F. Drivers shall bring their vehicles to a complete stop at all unguarded railroad grade crossings and shall not cross the tracks until determining that it is safe to do so.

IV. Responsibilities

A. Drivers shall be directly responsible for the safe and prudent operation of their vehicles in all situations.

B. When a driver is under the direct supervision of an officer, the officer shall assume responsibility for the actions of the driver and shall be responsible for immediately correcting any unsafe condition.

Occupational Safety and Health, SOP 403.04

VEHICLE ACCIDENT REVIEW BOARD

I. **Scope**

This standard establishes the department's Vehicle Accident Review Board, hereafter referred to as the board. It was promulgated to:

A. Require that a systematic review of every accident involving a motorized vehicle be conducted.

B. Recommend to the fire chief corrective actions that may be taken to prevent vehicle accidents.

C. Recommend to the fire chief that disciplinary action be taken when appropriate.

II. **General**

A. The board shall conduct a standardized review of every accident involving a motorized vehicle owned or operated by the department.

B. The membership of the board shall consist of the following positions:

1. The fire department safety officer, who shall act as chair.
2. One member and one alternate from each of the following ranks: firefighter, driver, and captain. Each position shall be elected by the members holding the same rank.
3. A nonsworn (or civilian or administrative) member and one alternate shall be elected by the nonsworn members of the department.

C. Members elected to serve on the board shall serve a term of two years.

D. To maintain consistency and continuity on the board, the nonsworn member, the driver, and their respective alternates shall be elected during odd years.

E. The firefighter, captain, and their respective alternates shall be elected during even years.

F. The election to select board members shall be conducted during the annual vacation pick each year (December 1, 2, and 3).

III. **Procedure**

A. Every vehicle accident shall be reviewed using the board guidelines listed below. A written report detailing the background of the accident, board results, and a recommended action shall be forwarded to the fire chief for each incident.

B. A copy of the board's report shall be forwarded to the Risk Management Office for analysis and review.

C. A copy of the board's report shall also be provided to each member involved in the accident, and a copy of the report shall be placed in the member's permanent personnel file.

D. Disciplinary action shall be administered as provided in the appropriate section of the department's rules, regulations, and procedures.

IV. **Guidelines**

A. All accidents shall fall into one of three categories:

1. Category One: Nonpreventable.
2. Category Two: Driver partially at fault.
3. Category Three: Driver totally at fault.

B. Category One accidents are those accidents in which no action could have been taken by the driver to prevent the accident. The board shall require that the appropriate written reports be filed and no disciplinary action be taken.
C. Category Two accidents are those accidents in which the driver is judged to have been partially at fault and could have prevented the accident. Disciplinary action may range from an oral reprimand for a first offense where the total physical damage was no more than $250.00 to a three-day suspension without pay when the driver has been involved in a preventable accident within the previous 24 months and the physical damage exceeded $250.00.
D. Category Three accidents are those accidents in which the driver was totally at fault. Disciplinary action shall be in direct proportion to the seriousness of the accident and shall range from a written reprimand to an indefinite suspension.

   Examples:
   1. A written reprimand should be sufficient for a first offense when the damage was less than $1,000.00 and there were no deaths or bodily injuries.
   2. An indefinite suspension should be recommended when one or more of the following conditions were present:
      a. The driver had multiple offenses during the previous 24 months.
      b. The damage exceeded $5,000.00.
      c. There was bodily injury or death.
      d. The driver was intoxicated or otherwise impaired.

Occupational Safety and Health, SOP 404.01

FIRE STATION SAFETY

I. **Scope**

This standard establishes safety regulations to be followed by members assigned to a fire station.

II. **General**

A. Fire stations shall comply with all applicable health, safety, building, and fire code requirements.
B. All fire stations are designated as tobacco free, and no one will be permitted to smoke or use smokeless tobacco products within a fire station. The use of tobacco will be confined to areas outside of the building.
C. Floors shall be kept clean and free from obstruction. Slippery substances such as water, oil, and other fluids shall not be allowed to accumulate on a floor surface and shall be mopped up as soon as is practical.
D. All tools and equipment shall be maintained in a clean and serviceable condition and shall be returned to their proper place immediately after use.
E. All flammable and combustible liquids and gases shall be stored in the station's flammable liquids cabinet. The cabinet shall be maintained in a clean and orderly manner and shall be kept closed and free of obstructions.
F. Prior to each use, all electrical equipment such as extension cords shall be inspected to prevent the possibility of shock or electrocution.

G. Horseplay is strictly forbidden.
H. Proper care shall be exercised when using any chemical product, pesticide, solvent, or other harmful or toxic substance.
I. Caution shall be exercised when using a ladder for cleaning, painting, etc.
J. All smoke detectors, fire extinguishers, exit signs, and other safety equipment shall be maintained in proper working order.
K. Running inside the station is prohibited.
L. Any defective equipment or unsafe condition shall be reported immediately.
M. Fire stations shall always be locked and secured whenever the station is unattended.
N. Automatic overhead door closures shall not be activated until the apparatus has completely cleared the door.
O. All lawn work shall be conducted in proper attire and the appropriate safety precautions shall be taken.
P. Apparatus will not be run in the engine bays unless the doors can be opened to allow the removal of engine exhaust or the apparatus' exhaust is connected to the exhaust removal system.

III. Responsibilities

A. Members shall strictly adhere to all safety regulations.
B. Officers shall be responsible for maintaining their assigned station in a safe and healthy manner and shall promptly correct any deficiencies.
C. Any member who violates a safety regulation shall be promptly reprimanded and the violation shall be reported to the fire chief.
D. The safety officer and a member of the Fire Prevention Bureau shall inspect each fire station at least once during each six-month period. The inspectors shall check the station for compliance with all applicable codes and safety standards. The inspection shall be conducted using the department's Inspection Report Form. A copy of the form will be given to the officer on duty at the time of the inspection and a copy shall be forwarded to the fire chief.
   1. If possible, corrections should be made immediately.
   2. If necessary, a reinspection shall be scheduled prior to concluding the inspection.
   3. The results of these inspections shall be reviewed during the officer's annual performance evaluation.

Occupational Safety and Health, SOP 405.01

PROTECTIVE CLOTHING AND EQUIPMENT

I. Scope

This standard applies to all members required to work in hazardous environments. It was promulgated to establish guidelines for the use of protective clothing and equipment to reduce the risk of illness, injury, or death that might result from a member's exposure to a hazardous environment.

II. General

A. Each member shall wear protective clothing and use equipment appropriate for the hazards to which he is exposed.

B. Each member shall properly maintain the protective clothing and equipment that have been issued to him and that are carried on the apparatus to which he has been assigned.

C. A standard washing machine and dryer have been provided at each station to allow members to maintain their work uniforms.

D. A protective clothing washer has been installed at Station _ to assist members in maintaining their turnout coat and pants. Each member shall clean his structural protective clothing at least once every six months and every time it is contaminated. (Local jurisdictions may make other provisions to accomplish this task, such as using a commercial service.)

E. Any equipment or protective clothing that is found to be unsafe or inoperable shall be red-tagged, removed from service immediately, and forwarded to the appropriate officer for repair.

III. **Structural Firefighting**

A. Each member assigned to fight structure fires shall be issued protective clothing and equipment that comply with all applicable NFPA standards and local requirements. These items shall include:

1. Helmet (NFPA 1972).
2. Boots (NFPA 1974).
3. Gloves (NFPA 1973).
4. Coat (NFPA 1971).
5. Pants with suspenders (NFPA 1971).
6. SCBA facepiece (NFPA 1981).
7. Hood (NFPA 1971).
8. Work uniform (NFPA 1975).

B. No alterations shall be made to protective clothing without the approval of the fire chief.

C. Each member operating within a perimeter designated as hazardous by the incident safety officer shall wear his full protective clothing in the prescribed manner. The incident safety officer shall determine when and if it is safe to remove some or all of the clothing.

D. Members shall be fully clothed beneath their turnouts (wearing the appropriate work uniform).

E. To assist members with identifying rank at an incident scene, helmets are color-coded as follows:

1. White: chief officers.
2. Blue: fire marshal's office.
3. Red: company officers.
4. Yellow: drivers/firefighters.

IV. **SCBA**

A. All self-contained breathing apparatus used by the department shall comply with NFPA 1981.

B. Unless the safety of the atmosphere can be determined by testing and continuous monitoring, all personnel shall use SCBA while working in areas where:

1. The atmosphere is hazardous.
2. The atmosphere is suspected of being hazardous.

3. The atmosphere may rapidly become hazardous.

C. Members wearing SCBA shall always work in teams of at least two members each.

D. SCBA and spare cylinders shall be kept on each apparatus and be available for immediate use.

E. SCBA shall not be removed until the incident safety officer has determined by testing that the atmosphere is no longer hazardous and that CO levels are less than 50 ppm.

F. A personal alert safety system (PASS) device that complies with NFPA 1982 shall be assigned to each SCBA and shall be activated whenever the SCBA is in use.

V. Wildland Firefighting

A. Protective clothing and equipment that meet NFPA 1977, *Standard on Protective Clothing and Equipment for Wildland Fire Fighting,* shall be provided for use by each member assigned to a brush company. This includes coveralls or pants and shirts, gloves, footwear, helmets, and eye protection.

B. Wildland gear shall be worn in lieu of structural protective clothing when fighting grass and brush fires whenever it is available.

C. If wildland gear is unavailable, the incident safety officer shall determine the level of protection that is required. At minimum, protection shall include Nomex® pants and shirt, gloves, helmet, eye protection, and footwear that meets NFPA 1974 or NFPA 1977.

VI. Life Safety Ropes, Harnesses, and Hardware

A. All life safety ropes, harnesses, and hardware used by the department shall meet the applicable requirements of NFPA 1983, *Standard on Fire Service Life Safety Rope and System Components.*

B. Class I life safety harnesses shall only be used for firefighter attachment to ladders and aerial devices.

C. Class II and Class III life safety harnesses shall be used for fall arrest and rappeling operations.

D. Rope used to support the weight of members or other persons during rescue, firefighting, other emergency operations, or training evolutions shall be life safety rope. Life safety rope used for any other purpose shall be removed from service and destroyed.

E. Life safety rope used for rescue at fires or other emergency incidents or for training shall be inspected before and after each use in accordance with the manufacturer's instructions and may be reused provided it has not sustained any visual damage due to heat, direct flame impingement, chemical exposure, or abrasion.

F. Life safety rope shall be removed from service and destroyed if the rope has been subjected to an impact load or exposure to a chemical known to deteriorate rope.

VII. Other Protective Clothing and Equipment

A. Emergency medical incidents: See SOP 702.03, Protective Clothing.

B. Hearing protection: See SOP 405.02, Hearing Conservation.

C. Water rescues: See SOP 602.01, Water Rescue.

VIII. Responsibilities

A. Each member shall inspect protective clothing and equipment issued to him and assigned to his apparatus at the beginning of each shift. Each member is responsible for the cleaning, care, and maintenance of his clothing and equipment and for obtaining repairs or replacement items.

B. Officers shall inspect protective clothing and equipment issued to the members under their command during the first shift of each month to ensure that all items are being properly maintained and are functioning properly.

C. The incident safety officer shall be responsible for ensuring that all members operating within the hazardous perimeter of an incident are properly attired and using the appropriate protective clothing and equipment. Any member who is in violation of this procedure shall be ordered to promptly leave the hazardous area.

D. A member who uses a life safety rope shall have the rope inspected by the small-equipment officer before the rope can be returned to service.

E. The small-equipment officer shall maintain a record of all rope use and shall remove from service all rope deemed to be unsafe.

Occupational Safety and Health, SOP 405.02

HEARING CONSERVATION

I. Scope

This standard establishes a hearing conservation program for the department. It was promulgated to prevent job-related hearing impairment.

II. Audiometric Testing

A. As a part of the entry level medical examination, a member shall be required to have a baseline audiogram performed. This baseline audiogram will become part of the member's permanent medical history file and will be used throughout his tenure with the department to monitor any changes in his hearing.

B. An audiometric test shall also be conducted as a part of the annual medical evaluation, and the results will be included in the member's medical history file.

III. Hearing Protection

A. The noise level will be monitored whenever a process or equipment change occurs.

B. Warning signs will be posted at the entrances to or on the periphery of work areas where a member may be exposed to a sound level of 90 dBA or greater.

C. Warning signs will clearly indicate that the area is a noise hazard area and that hearing protection must be worn while in the area.

D. Hearing protection shall be mandatory in all areas marked as hearing protection areas.

E. Hearing protection shall also be worn when noisy jobs or tasks are being performed in an area not posted as a hearing protection area, such as:

1. While performing noisy tasks such as grinding, operating air tools and fire pumps, etc.

2. Whenever an employee has to raise his voice to be heard.

F. Whenever possible, high noise exposures will be reduced to acceptable levels by using proper administrative or engineering controls.

G. Personal protective equipment will be provided and worn by members when administrative or engineering controls are not feasible or prove ineffective in reducing high noise exposures to acceptable levels.

IV. Responsibilities

A. The safety officer shall be responsible for the identification of products, systems, or operations where the noise level exceeds 90 dBA and for developing control procedures to mitigate the hazard.

B. Officers shall be responsible for enforcing the provisions of this standard and for immediately correcting any deficiencies that might occur.

C. Each member shall be responsible for strictly adhering to the provisions of this standard and for wearing an approved hearing protection device whenever a hearing protection sign is displayed or whenever engaged in an activity where the noise level exceeds 90 dBA.

Occupational Safety and Health, SOP 405.03

FLASHOVER/BACKDRAFT REPORT

I. Scope

This standard applies to incidents where a flashover or backdraft occurs. It was promulgated to provide a database for research into the phenomena of flashovers and backdrafts.

II. General

A. Backdrafts, flashovers, and similar phenomena are extremely hazardous to firefighters. In an effort to better understand the dynamics of these occurrences, the incident commander shall complete a Flashover/ Backdraft Report whenever a flashover or backdraft occurs.

B. The fire chief shall maintain a file on the completed reports. The reports will be used for research and analysis. Hopefully, this research will result in the reduction of injuries and deaths to firefighters.

Occupational Safety and Health, SOP 405.04

REHABILITATION

I. Scope

This standard applies to all emergency operations and training exercises where strenuous physical activity or exposure to heat or cold creates the need for the rehabilitation of personnel. It was promulgated to:

A. Prevent injuries, illnesses, and deaths that may result from excessive fatigue.

B. Establish procedures for medical evaluation and treatment, food and fluid replenishment, and relief from extreme climatic conditions during emergency operations and prolonged training exercises.

II. Definitions

A. Level I rehabilitation: Situations of short duration. The incident commander may elect to use the rehabilitation supplies from an apparatus on the scene or may special-call the rehab unit to the scene. Typically in Level I rehab, the crews are not rotated and the incident or training exercise has a limited duration.

B. Level II rehabilitation: Situations that require a major time and personnel commitment. Examples include a major fire or a lengthy training exercise in which the firefighter's health and safety must be addressed.

III. General

A. The incident commander shall evaluate the circumstances at each incident and shall make early, adequate provisions for the rest and rehab of all members working at the scene. These provisions include:

1. Medical evaluation.
2. Treatment and monitoring.
3. Food and fluid replenishment.
4. Mental rest.
5. Relief from extreme climatic conditions and other environmental factors present at the incident.

B. During prolonged incidents, strenuous training sessions, and periods of extreme heat or cold, the incident commander shall request that the rehab unit be dispatched to the scene and shall appoint a rehab officer to manage the rehabilitation of the firefighters.

C. The Rehab Log shall be completed by the rehab officer at all Level II incidents. The log shall be submitted to the incident commander to be attached to the incident report, and it shall be included as part of the incident postmortem. Level II rehab includes the provision of EMS at the ALS level in the rehab sector.

IV. Rehabilitation Sector

A. The incident commander shall establish a rehabilitation sector when conditions indicate that members working at an incident or training exercise require rest and rehab.

B. The incident commander shall appoint a rehab officer who will assume command of the rehabilitation sector. At most incidents, the location of the rehabilitation sector will be designated by the incident commander. However, if the incident commander has not designated a rehab site prior to the appointment of a rehab officer, the rehab officer shall promptly select an appropriate location based on the most desirable site available.

C. The rehab sector should be placed in a location that allows members to physically and mentally rest and recuperate from the stress, pressure, and demands of the emergency operation or training evolution.

D. The rehab sector should also be located far enough away from the incident scene to allow members to safely remove their protective clothing and SCBA.

E. The rehab sector should be located in an area that provides suitable protection from the prevailing environmental conditions. If possible, the sector should be in a cool, shaded area during hot weather and in a warm, dry area during cold weather.

F. If the rehab sector is located outdoors, the area should be free of ants and other stinging or biting insects.

G. Members in the rehab sector should not be exposed to exhaust fumes from apparatus, vehicles, and motorized equipment, including those involved in the rehabilitation sector operations.

H. The rehab sector should be large enough to accommodate multiple crews and should allow for expansion or contraction as the size of the incident varies.

I. The rehab sector should be easily accessible by EMS units and other support vehicles.

J. The rehab sector should be located close enough to the incident to allow members to promptly reenter the emergency operation site after recuperation.

K. The following areas should be considered when selecting a site for the rehab sector:

1. A nearby garage, building lobby, or other structure.
2. At least two floors below a fire in a high-rise building.
3. A large climate-controlled vehicle such as a school or transit bus.
4. Fire apparatus, ambulances, or other emergency vehicles at the scene or called to the scene.
5. The fire department rehab unit.
6. An open area in which a rehab site can be created by using tarps, fans, etc.
7. At an industrial site, the rehab sector shall be placed *outside* the fenced compound area.

L. The rehab officer shall secure all the resources required to adequately staff and supply the rehab sector. The supplies may include the items listed below:

1. Fluids such as water, activity beverage, Gatorade®, and ice.
2. Food such as soup, broth, or stew in hot/cold cups.
3. Medical equipment such as blood pressure cuffs, stethoscopes, oxygen administration devices, cardiac monitors, intravenous solutions, and thermometers. (Medical supplies may be furnished by the ambulance assigned to the rehabilitation sector.)
4. Other items such as awnings, fans, tarps, smoke ejectors, heaters, dry clothing, extra equipment, floodlights, towels, traffic cones, and fire-line tape (to identify the entrance and exit of the rehabilitation area).

V. Guidelines

A. The establishment of a rehab sector shall be considered during the initial planning stages of an emergency response. The climatic or environmental conditions of the emergency scene should not be the sole justification for establishing a rehab sector. Any activity or incident, whether emergency or nonemergency, that is large in size, long in duration, and labor intensive will rapidly deplete the energy and strength of personnel.

B. Climatic or environmental conditions that indicate the need to establish a Rehabilitation Sector include a heat index above 95°F or a windchill index below 20°F.

C. A critical factor in the prevention of heat stress injury is the intake of water and electrolytes during periods of intense physical activity.

1. During these periods, an individual should drink at least one quart of water or Gatorade® per hour.

2. Adequate fluid intake is important even during cold weather operations. Despite outside temperatures, heat stress injuries may occur during firefighting or other strenuous activity anytime that protective clothing and equipment are worn.
3. Individuals should avoid caffeinated and carbonated beverages because both interfere with the body's water conservation mechanisms.
4. Certain drugs also impair the body's ability to sweat. Use caution if a member has taken antihistamines, diuretics, or stimulants.

D. If the duration of an incident extends through regular mealtimes, the department shall provide food to the members at the scene whenever it is possible to do so. Food may be charged to the department at certain stores. The department may also have a canteen or use the Red Cross, Salvation Army, or local buff or auxiliary group.

E. Forty-five minutes of work time is generally recommended as an acceptable level of work prior to mandatory rehabilitation.
1. Members having worked through two full 30-minute-rated SCBA cylinders, or for 45 minutes, shall be rotated to the rehabilitation sector for rest and evaluation.
2. In all cases, an objective evaluation of a member's fatigue level is the appropriate criterion for determining if rest is required. Rest periods for members in the rehab sector shall be at least 10 minutes or greater.
3. Crews sent to rehab should be replaced by fresh crews from the staging sector. Crews released from the rehab sector should be rotated to the staging sector prior to returning to work. This procedure ensures that fatigued individuals do not return to work before they are rehabilitated.

F. Members should not be removed from a hot environment and placed directly into an air-conditioned environment because the body's cooling system may shut down in response to the external cooling. An air-conditioned environment is acceptable only after a cooldown period at ambient temperature with sufficient air movement.

G. EMS at the advanced life support level will be provided at each incident. EMS personnel (paramedics) will evaluate the vital signs and the physical condition of members as they are rotated through the rehab sector.
1. EMS personnel will determine whether a member will be allowed to return to work, remain in rehab, or receive further medical treatment and be transported to a medical facility for further evaluation.
2. Continued rehabilitation consists of the ongoing monitoring of vital signs, rest, and fluid intake. Medical treatment for a member whose signs and symptoms indicate potential problems will be provided in accordance with local medical control procedures. EMS personnel will be aggressive in determining that potential medical problems exist.

H. When working crews arrive at the rehab sector, each member's vital signs shall be taken and recorded. The following criteria are used in the evaluation of fireground personnel during a fire or EMS incident:
1. Transportation to the hospital is required when the diastolic blood pressure is ≥ 130.

2. Transportation to the hospital is also required when the diastolic blood pressure is ≥ 110 and the individual is symptomatic.
3. An individual may be transported to the hospital for further evaluation when the diastolic blood pressure is ≤ 110 and the individual is symptomatic.
4. The individual may be transported when the systolic blood pressure is ≥ 200 and after further evaluation and rest the systolic blood pressure is still ≥ 200.
5. When a pulse rate of ≥ 140 is found, administer oxygen and fluids, rest for a minimum of 10 minutes, and reassess the individual. If the heart rate is less than 140, the individual may return to work.
6. If after 10 minutes the heart rate still remains above 140, the individual must rest for an additional 30 minutes. Administer fluids and oxygen, and record the heart rate and rhythm on a cardiac monitor and obtain an EKG printout.
7. If after 30 minutes the pulse rate remains above 140, transport the member to a medical facility for further evaluation.
8. In the above cases, Medical Control will be contacted in every situation and treatment or transport will be determined in conjunction with Medical Control.

I. All medical evaluations shall be properly recorded by the paramedic, along with the individual's name and chief complaints. The form must list the date, time, and incident number and be signed by the rehab officer.

J. Members sent to rehab shall enter and exit the rehab sector as a crew. The crew designation, number of crew members, and times of entry to and exit from the rehab sector shall be documented by the rehab officer. Crews shall not leave the rehab sector until released by the rehab officer.

VI. Maintenance of the Rehab Unit

A. The rehab unit shall be checked at the beginning of each shift.

B. The rehab unit shall also be checked and cleaned after each use by the members who used the vehicle. The Rehab Unit Checklist shall be completed and turned in with the regular paperwork each time the rehab unit is used. The rehab unit is not kept fully stocked at all times. Before responding with the rehab vehicle, a member should check to determine whether drinks and candy have been restocked.

VII. Responsibilities

A. All officers shall monitor the condition of each member working under their command and shall ensure that adequate steps are taken to provide for each member's safety and health. The incident command system is to be used to request that a crew be relieved and for the reassignment of fatigued crews.

B. During periods of hot weather, members are encouraged to use their individual water bottles and drink water or Gatorade® throughout the workday. During any emergency incident or training evolution, all members shall advise their supervisor when they believe their level of fatigue or exposure to heat or cold is approaching a point that could affect them, their crew, or the operation in which they are involved. Each member shall also monitor the health and safety of the other members of his crew.

VIII. **Heat Stress Index**

TEMPERATURE °F	DANGER	INJURY THREAT CATEGORY
Below 80°F	None	Little or no danger under normal circumstances.
80°F-90°F	Caution	Fatigue possible if exposure is prolonged and there is physical activity.
90°F-105°F	High	Heat cramps and heat exhaustion possible if exposure is prolonged and there is physical activity.
105°F-130°F	Extreme	Heat cramps or exhaustion likely; heat stroke possible if exposure is prolonged and there is physical activity.
Above 130°F	Mortal	Heat stroke imminent!

Add 10° F when protective clothing is worn and add 10° F. when in direct sunlight.

IX. **Windchill Index**

	WINDCHILL TEMPERATURE (°F)	DANGER
A	Above 25°F—Little danger for properly clothed person.	
B	25°F to -75°F—Increasing danger, flesh may freeze.	
C	Below -75°F—Great danger, flesh may freeze in 30 seconds.	

Occupational Safety and Health, SOP 405.05

OPERATING AT EMERGENCY INCIDENTS

I. **Scope**

This standard applies to members operating at an emergency incident. It was promulgated to:

A. Prevent accidents, injuries, and deaths that might result from an unsafe act while members are operating at an emergency incident.

B. Define the minimum personnel requirements for the safe conduct of emergency scene operations.

II. **Guidelines**

A. No member shall commence or perform any firefighting or rescue function or evolution that is not within the established safety criteria of the department.

1. Activities that present a significant risk to the safety of a member shall be limited to situations where there is a potential to save endangered lives.
2. It is unacceptable to risk the safety of a member when there is no chance of saving lives or property.
3. In situations where the risk to a member is excessive, activities shall be limited to defensive operations.

B. When an inexperienced member is working at an incident, direct supervision by an experienced officer or member shall be provided.

C. Members operating in hazardous areas at emergency incidents shall operate in teams of two or more. Team members operating in hazardous areas shall be in constant communication with each other through visual, auditory, or physical means or through the use of a safety guide rope so as to coordinate their activities. Team members shall remain in close proximity to each other to provide assistance in case of emergency.

D. An interior firefighting effort at a working structural fire shall not take place until a minimum of (recommended: four) _____ firefighters are present.

1. When only four firefighters are present, two members shall work as a team in the hazardous area and two members shall remain outside the hazardous area and be available for entry into the hazardous area if assistance or rescue is required.
2. A working fire is defined as a fire that requires the use of a 1½-inch or larger attack line and the use of SCBA.

E. The standby members shall be responsible for maintaining a constant awareness of the number and identity of the members operating in the hazardous area, their location and function, and their time of entry. The standby members shall remain in radio, visual, voice, or signal line communication with the team.

F. One standby member shall be permitted to perform other duties outside of the hazardous area, such as serving as an apparatus operator or incident commander, provided that constant communication is maintained between the standby member and the members of the team in the hazardous area.

1. The assignment of personnel to other duties shall not be permitted if the abandonment of their assignment would jeopardize the safety and health of any firefighter working at the incident.
2. The assignment of personnel to other duties shall not be permitted if their assignment inhibits their ability to assist in or perform a rescue.
3. Standby members shall have full protective clothing and SCBA available to them as defined by SOP 405.01, Protective Clothing and Equipment.

G. Exception: Rescue operations may be undertaken prior to the assembly of four firefighters if there is an imminent life-threatening situation and immediate action could prevent the loss of life or serious injury. No exception shall be permitted when there is no chance of saving lives.

H. When a second team is assigned to or begins operating in the hazardous area, the incident commander shall designate at least one rapid intervention team (RIT) to stand by in the event that a rescue becomes necessary.

1. The RIT shall consist of at least two members. The team shall have full protective clothing and SCBA available to them as defined by SOP 405.01, Protective Clothing and Equipment.
2. During the initial stages of an incident, the RIT may be used to

perform other functions provided that the team is immediately available to perform a rescue if so required.

3. As an incident grows in complexity, the number of RITs shall be increased proportionately and shall be dedicated solely to this responsibility.

I. At least one ambulance shall stand by during all working incidents and be available to treat injuries and to provide transport if necessary.

J. Members operating from aerial devices shall be secured to the device by an approved safety harness.

K. Apparatus shall be used as a shield against oncoming traffic wherever possible.

III. Emergency Communications

A. RITs shall be provided with portable radios and shall monitor the fireground frequency.

B. The term *Mayday* shall be used by anyone on the scene who becomes aware of or is involved in a life-threatening situation.

C. The term *emergency* shall be used by anyone on the scene who needs to communicate an urgent message.

D. During Mayday or emergency traffic conditions, all other incident radio traffic shall immediately stop.

IV. Responsibilities

A. It shall be the responsibility of each member to fully comply with the provisions of this standard.

B. Officers shall be responsible for keeping their crews together and for ensuring that they do not expose their crews to unnecessary risks.

C. The incident commander shall be responsible for ordering sufficient resources to ensure that all emergency incident functions are performed in a safe manner.

Occupational Safety and Health, SOP 405.06

ACCOUNTABILITY

I. Scope

This standard applies to members operating at an emergency incident. It was promulgated to provide a structured approach for tracking all members operating at an emergency scene.

II. Guidelines

A. The provisions of this standard shall be followed whenever members are required to work in an environment that requires the use of SCBA or where a member may become lost, trapped, or injured by the environment.

B. Members operating in a hazardous environment as defined above shall maintain company or crew integrity and shall use the buddy system. This requires that:

1. Company or crew members enter and exit the environment together.
2. Members remain within either sight, voice, or tactile distance of each other while they are within the environment. No one shall ever be left alone.

3. Incident commanders and sector officers shall not direct members to operate independently of their companies or crews.
4. Task assignments shall be made through the company officer or crew leader.

C. A company or crew may be divided into multiple teams to perform tasks that do not require the efforts of the entire company, provided that:

1. Each team has a minimum of two members.
2. Each team is equipped with a portable radio.
3. The officer remains in contact with each team and is constantly aware of their locations.
4. A given team enters and exits the hazardous environment together.
5. On completion of an assigned task, the team leader reports to the company officer or crew leader for another assignment.

III. **Accountability Equipment**

Note: A number of excellent accountability systems are available. The system described below is just one of them. If your department is regularly involved in automatic or mutual aid, adopt a system that is compatible with your neighbor's.

A. Helmet identification tag: A thin metal plate that identifies the company or crew to which a member is assigned is attached to each side of the helmet by a self-adhesive fastener.

1. Helmet tags shall be exchanged at shift change by arriving personnel.
2. Extra sets of tags are provided to each company for quick replacement of a lost or damaged tag.

B. Personnel accountability tag: A plastic tag that is engraved with the member's name and identification number. Every member assigned to emergency response duty should be issued a personnel accountability tag.

1. When the member is off duty, the tag is to be placed on the top snap hook of the member's turnout coat.
2. When the member is on duty, the tag shall be placed on the company responder board.
3. At shift change, an arriving member shall remove the tag of the member he is relieving from the responder board and place the tag on the top snap hook of the member's turnout coat. The arriving member shall then place his own tag on the company responder board in the appropriate position.

C. Company responder board: A board mounted on the dash of each apparatus in full view of the company officer or crew leader. The board shall remain with the apparatus at all times unless Command orders it to be removed to locate a lost member.

1. Members shall attach and remove their tags as appropriate.
2. The company officer or crew leader shall update the company responder board as required and remove the tag of any member not on board the apparatus.

D. Passport: A metal tag labeled with the company's identity that is used to account for each company operating at an incident.

1. Company officers or crew leaders shall transfer their passports to the incident commander or a sector officer at their first face-to-face encounter.
2. The incident commander shall hold the passports for companies and sector officers directly under their span of control.
3. Sector officers shall hold the passports for companies directly under their span of control.
4. If a company is transferred to a different sector, the company officer shall retrieve its passport and transfer it to the new sector officer.
5. If a company is forced to exit at a location other than its original point of entry, it shall immediately notify its sector officer and make every effort to retrieve its passport.
6. When it is no longer necessary to use the passports, the incident commander shall announce "Store passports," and the company officers shall retrieve their passports at the first opportunity.

E. Makeup set: A blank responder board, personnel accountability tags, and a passport. A set is carried on each apparatus for use by a mutual aid company that does not have accountability equipment.
   1. The mutual aid company's makeup company responder board shall be given to the incident commander prior to receiving an assignment.
   2. Makeup sets may also be used by off-duty personnel as necessary in the event that they arrive at the incident without their own personnel accountability tag.

IV. Roll Call

A. A roll call is a systematic method for reporting to command that all members operating at an incident are accounted for. It should be conducted periodically throughout an incident to ensure that all members are safe and accounted for.

B. A roll call shall be conducted:
   1. When changing from an offensive to a defensive mode.
   2. When an unexpected or catastrophic event occurs, for example in the case of flashover, backdraft, or structural collapse.
   3. After an emergency evacuation.
   4. At the first report that a member is missing.
   5. When a fire is declared to be under control.
   6. Prior to suspending the use of passports.
   7. At the discretion of Command. A localized report may also be conducted by a sector officer.

C. On receipt of an order to conduct a roll call:
   1. Company officers or crew leaders shall confirm that their personnel are accounted for and shall notify their sector officer.
   2. Sector officers shall notify Command when all of the members under their span of control have been accounted for.
   3. The roll call is complete when everyone has been properly accounted for. This includes staff support assigned to a sector or to Command.
   4. Dispatch shall repeat the announcement and give the correct time. The incident commander shall record the time on the incident worksheet.

5. Unless directed otherwise, operations will continue while the report is being taken.

V. **Lost or Trapped Member**

A. In the event that one or more members cannot be accounted for, a roll call shall be requested immediately by the member or officer who believes that a member is missing.

B. If a member cannot be accounted for during a roll call, he will be presumed lost until he can be accounted for.

C. An attempted rescue will become the top priority at the incident and sufficient resources shall immediately be assigned to conduct the rescue effort.

D. Command shall retrieve the responder board from the missing member's company to identify the missing member.

E. As soon as all members have been positively accounted for, the rescue effort shall be suspended.

VI. **Emergency Communications**

A. The term *emergency* shall be used by anyone on the scene who needs to communicate an urgent message.

B. The term *Mayday* shall be used by anyone on the scene who becomes aware of or is involved in a life-threatening situation.

VII. **Emergency Evacuation**

A. When it is unsafe to continue emergency operations, Command shall order a rapid and complete evacuation from the hazardous environment.

B. An evacuation may be initiated by anyone at the incident, but the order to evacuate must be transmitted through the established chain of command to ensure that everyone can be accounted for.

C. The order to evacuate shall be transmitted as follows: "Interior Sector to Command, emergency! Evacuate the building immediately!"

D. Command shall immediately order everyone to evacuate the hazardous environment. Example: "Maple Street Command, emergency! All units evacuate the building immediately! Repeat, all units evacuate the building immediately!"

E. All companies and each sector shall acknowledge the order to evacuate.

F. Command shall conduct a roll call after everyone has acknowledged the order to evacuate.

Occupational Safety and Health, SOP 406.01

FITNESS FOR DUTY

I. **Scope**

This standard shall apply to all members required to engage in emergency operations. It was promulgated to:

A. Evaluate the medical and physical fitness and ability of members engaged in emergency operations.

B. Certify that members engaged in emergency operations are medically and physically fit and able to perform their duties.

II. **Annual Medical Evaluation**

A. All members assigned to emergency operations duty shall be annually certified by the department's physician as meeting the medical requirements of NFPA 1582, *Standard on Medical Requirements for Fire Fighters*.

B. The annual evaluation shall be scheduled within 30 days of a given member's anniversary date.

C. The annual medical evaluation shall consist of:

1. An interval medical history.
2. An interval occupational history, including significant exposures.
3. Height and weight.
4. Blood pressure.

D. The annual medical examination shall be conducted as follows:

1. Ages 29 and under—every three years.
2. Ages 30 to 39—every two years.
3. Ages 40 and above—every year.

E. Any member who is not certified as meeting the medical requirements of NFPA 1582 shall be relieved of emergency operations duty and shall be assigned to other duties by the fire chief until such time as the member can be certified to return to emergency operations duty by the department's physician. The department's physician shall place the member in a rehabilitation program to prepare him to return to emergency duty.

F. If any member is unable to return to full duty within 12 months of being relieved of duty, the member shall be reevaluated and a decision made as to his future with the department.

G. If a member has an acute illness or other condition that prevents him from being evaluated at his normally scheduled time, the evaluation shall be postponed until he has sufficiently recovered.

H. Any member who has been absent from duty for a medical condition, injury, or other reason that may affect his performance may not return to duty until certified fit for duty by the department's physician.

III. **Physical Fitness**

A. Each member assigned to emergency operations shall maintain his personal physical fitness at a level sufficient to pass the annual physical performance examination.

B. All members shall participate in a physical fitness activity during their assigned workout period whenever response activity allows. To assist in this process, a variety of physical conditioning equipment shall be provided at each station. In addition, workout periods shall be assigned to each company.

C. If a member is unable to participate, he shall be sent home on sick leave. If a member refuses to participate, he shall be placed on administrative leave and charged with insubordination by his supervisor.

IV. **Annual Physical Performance Evaluation**

Note: This section is a recommendation. Local practices may vary.

A. Each member shall be evaluated and certified annually as meeting the department's minimum physical performance requirements.

B. The annual evaluation shall be scheduled and conducted by the wellness officer during a set period each year.

C. The evaluation shall consist of passing the department's entry-level physical ability test, except that the test shall be conducted while the member is wearing full structural protective clothing.

D. Members who fail to meet the minimum physical performance requirements shall be placed in a rehabilitation program, not to exceed 120 days, to facilitate their progress in attaining the minimum level of performance required.

E. If after 120 days a member is still unable to meet the minimum physical performance requirements, he shall be assigned to other duties by the fire chief. The member shall then have 12 months to attain the required level. If he is still unable to attain the minimum required level, his future with the department shall be evaluated.

Occupational Safety and Health, SOP 406.02

PERMANENT MEDICAL FILE

I. Scope

This standard shall apply to all members of the department. It was promulgated to establish the requirement that a permanent, confidential medical file be created and maintained for each member of the department.

II. General

A. A permanent, confidential medical file shall be maintained by the department for each member and shall be periodically updated during the member's tenure with the department.

B. The file shall contain:

1. A copy of the initial or baseline medical examination performed at the time an individual was accepted as a member of the department and/or the initiation of the wellness program. (Note: Some departments may not require a medical examination at the time a person becomes a member. If a wellness program is established, a baseline physical should be performed for each member.)
2. Copies of the results of all subsequent medical evaluations and physical performance tests.
3. Records of any occupational illness or injury.
4. Copies of all reports of exposures to hazardous materials and contagious diseases. See SOP 406.03, Exposure Reporting.
5. A copy of the autopsy results in the event of death due to an occupational illness or injury.

C. The medical file is strictly confidential. Only the member, his physician, the department's physician, and the fire chief or his designee shall be granted access to this file.

Occupational Safety and Health, SOP 406.03

EXPOSURE REPORTING

I. Scope

This standard shall apply to any member who has been exposed to or has been in contact with any hazardous material or contagious disease while on duty. It was promulgated to:

A. Provide a history of each member's exposure to hazardous materials and contagious diseases while on duty.

B. Identify the long-term health problems associated with these exposures.

II. Procedure

A. Whenever a member is exposed to or comes in contact with a hazardous material or contagious disease while on duty, he shall complete a record of the incident on the Hazardous Materials Exposure Form. The completed form shall be forwarded to the fire chief within 24 hours of the exposure.

B. A copy of the completed form shall be placed in the member's permanent medical file.

C. Coding instructions:

1. Insert the member's name and SSN, the incident number, and the date of exposure.
2. List the type of incident:
    a. Chemical reaction.
    b. Spill or leak.
    c. Fire.
    d. Explosion.
    e. Vapor release.
    f. Medical call.
    g. Any combination of the above.
3. Level of protection used:
    a. Full structural protective clothing, SCBA, and duct tape for all exposed areas.
    b. Full-face cartridge respirator with the appropriate cartridge, hard hat, appropriate splash clothing, gloves, and boots.
    c. Fully encapsulated suit and SCBA.
    d. Thermal protection ensemble and SCBA.
4. Supplement number if more than two chemicals are encountered by the individual.
5. Insert the time and location of the incident.
6. Vital signs taken on the initial medical survey. Each member who is directly involved in the operation shall have a set of vitals taken.
7. Length of the exposure in minutes:
    a. On the scene.
    b. In the hot zone.

8. Medical action:
   a. Conditions monitored on the scene by the paramedics—i.e., vitals, survey, etc.
   b. Transported to a medical facility for observation or treatment. Released with no further follow-up.
   c. Transported to a medical facility, laboratory work, and follow-up with a physician. Not admitted.
   d. Transported to a medical facility and admitted with ongoing observation or medical treatment with postrelease follow-up.
9. On-scene activity performed:
   a. Entered hot zone.
   b. Entered warm zone/decontamination.
   c. Remained in cold zone.
10. Hazard class:
    a. Explosives.
    b. Gases.
    c. Flammable and combustible liquids.
    d. Flammable solids.
    e. Oxidizers and organic peroxides.
    f. Poisons.
    g. Radioactive materials.
    h. Corrosives.
    i. ORM (other related materials).
    j. Biological.
11. United Nations Identification Number (DOT book).
12. Shipping name if known.
13. Trade name if known.
14. Biomedical name: Enter the biological or common medical name of the disease exposed to if known; attach a copy of the patient form to the Hazardous Materials Exposure Form.
15. Enter other pertinent information at the bottom of the page or on the back.

III. Responsibilities

A. Each member shall be responsible for completing the Hazardous Materials Exposure Form whenever he is exposed to or comes in contact with a hazardous material or contagious disease while on duty.

B. Officers shall be responsible for ensuring that members under their command have properly completed the Hazardous Materials Exposure Form whenever they have been exposed to or come in contact with a hazardous material or contagious disease. Officers shall forward the completed forms to the fire chief within 24 hours of the exposure.

Section 500

Maintenance

REPAIR REQUESTS

I. Scope

This standard establishes a procedure to identify items in need of repair and to request that repairs be performed. It was promulgated to:

A. Ensure the readiness of apparatus, small tools, and equipment by creating a process for identifying mechanical and electrical defects.

B. Establish a procedure for placing defective items out of service.

C. Establish a procedure for requesting that repairs be performed.

D. Assign individual responsibilities for the repair of defective items.

II. Repair Procedure

A. Apparatus, small tools, and equipment shall be periodically inspected to ascertain whether they are functioning properly or are in need of maintenance or repair. All inspections shall be conducted in accordance with the daily work schedule and applicable SOPs.

B. Whenever a defect or malfunction is discovered, the member who discovers it shall attempt to repair the defect provided that he has the appropriate tools, supplies, and expertise to do so.

C. If the member is unable to repair the item, an Equipment Service Request Form shall be completed for large pieces of equipment or a vehicle. The white copy shall be sent to the maintenance and logistics officer and the yellow copy shall be retained in the station log.

D. The completed Equipment Service Request Form shall then be forwarded to the member's officer. The officer shall note the problem in the Daily Log, sign the Equipment Service Request Form, and forward the form to the maintenance and logistics officer (MLO).

E. The MLO shall schedule the repairs and shall notify the officer when the repairs are expected to be completed.

F. If it is necessary to take an apparatus of service, follow the guidelines in SOP 501.03, Declaring a Vehicle Unsafe to Operate.

G. If it is necessary to take a small tool or a piece of equipment out of service, the item shall be tagged with a Red Tag Out-of-Service Card. The member shall record a description of the defect on the tag, the date the item was tagged, and the name of the member completing the tag.

H. Whenever possible, the item being taken out of service shall be replaced by a reserve item.

I. The red-tagged item shall be taken to the maintenance shop as soon as possible.

Maintenance, SOP 501.01

APPARATUS AND MOTORIZED VEHICLES

I. Scope

This standard establishes a schedule for the inspection and maintenance of all apparatus and motorized vehicles owned or operated by the department. It was promulgated to:

A. Ensure that emergency response vehicles are maintained in a constant state of readiness.

B. Implement a preventative maintenance schedule for all motorized vehicles.

C. Establish procedures for the daily inspection of apparatus, equipment, and support vehicles.

II. General

A. Apparatus and support vehicles shall be:

1. Kept clean at all times.
2. Maintained in a constant state of readiness.
3. Refueled whenever the fuel level drops below 3/4 of a tank. Oil and ancillary fluid reservoirs shall also be kept full at all times.

B. All engine-powered equipment shall be kept clean and their fuel tanks and oil and fluid reservoirs shall be refilled whenever the level drops below 3/4 of a tank.

III. Daily Inspections

A. Every vehicle garaged in a fire station shall be inspected by station personnel at the beginning of each shift. The member performing the inspection shall record his findings on the Driver's Daily Apparatus Checklist.

B. The member performing the inspection shall correct the defects that are found provided that the member has the expertise, tools, and supplies to do so. The items that are corrected shall be noted in the comments section of the checklist.

C. Defects that cannot immediately be corrected shall be noted on an Equipment Service Request Form.

D. Whenever a defect requires that a vehicle be placed out of service, the driver shall place the vehicle out of service in accordance with the provisions of SOP 501.03, Declaring a Vehicle Unsafe to Operate.

E. The completed checklists and service requests shall be forwarded to the officer for review and disposition. The officer shall review the work that was performed to ensure that the repairs have been satisfactorily made and shall review each checklist to ensure that it is complete and accurate.

F. The completed and signed checklists and service requests shall be forwarded to the maintenance and logistics officer (MLO) on a daily basis. The MLO shall schedule the repairs and shall notify the officer of when to deliver an apparatus or vehicle for maintenance.

IV. Weekly Maintenance

The following items shall be performed at least once each week as indicated on the Daily Work Schedule (See SOP 200.05, Daily Work Schedule). The officers shall note in their Daily Log the performance of all weekly maintenance activity.

A. Staff vehicles: Automobiles, pickups, and utility vehicles shall be checked and serviced at least once each week. The member assigned to perform the service shall complete a Small-Vehicle Weekly Checklist on each vehicle serviced and shall forward the completed form to the maintenance and logistics officer. In addition to inspecting the items on the checklist, each staff vehicle shall:
   1. Be thoroughly washed.
   2. Have its windows cleaned.
   3. Have its interior vacuumed and cleaned.

B. Aerial devices: The following tasks shall be performed on Tuesday of each week, after every major repair, and after each use:
   1. A visual inspection of all systems and components.
   2. The aerial shall be set up and operated to check the function of the outriggers, turntable, and aerial device.
   3. All defects shall be noted on the Equipment Service Request Form.

C. Reserve apparatus:
   1. A thorough inspection of each reserve piece of apparatus shall be made on Sundays, after any major repair, and after major use. Any item that needs repair shall be recorded on the Equipment Service Request Form.
   2. All reserve apparatus shall be driven at least once each week to ensure the proper function of all mechanical systems and to circulate all fluids.

D. Fire pumps:
   1. On Thursdays, after any major repair, and after each major use, the driver shall:
      a. Open all pump drains and flush out the sediment.
      b. Check and clean the intake strainers.
      c. Check the gearbox oil level.
      d. Operate the pump primer with all valves closed.
      e. Operate the transfer valve while pumping from the booster tank.
      f. Check the pump seals for leaks.
      g. Operate all valves.
      h. Operate the relief valve.
      i. Check all gauges and flow meters for proper operation.
   2. The driver shall fill out an Equipment Service Request Form and list any items that do not function properly.

V. Quarterly Maintenance

A. In addition to routine daily maintenance, apparatus and motor vehicles shall undergo the quarterly maintenance listed on the Preventive Maintenance Worksheet.

B. Quarterly maintenance shall be completed on the first day of January, April, July, and October and shall include:
   1. A complete degreasing. This includes a hot-water wash of the

undercarriage, frame, axles, motor, pumps, spring shackles, tie rod ends, and turntable assemblies.

2. The motor oil shall be changed and the vehicle shall be lubricated.
3. All work performed during the quarterly maintenance cycle shall be noted on the checklist along with the amount of parts and supplies used and the personnel hours required to complete the task. The completed form shall be reviewed and signed by the station officer and then forwarded to the maintenance and logistics officer.
4. Any items in need of repair shall be recorded on the Equipment Service Request Form.

C. Officers shall record the performance of any quarterly maintenance activity in their Daily Log.

VI. **Semiannual Maintenance**

A. The following items shall be performed during the first week of March and September:

1. All apparatus and support vehicles shall be washed, compounded, and waxed.
2. The motor oil and filter on the drive engine shall be changed. The fluid levels of the transmission, pump transmission, differential, power steering, primer reservoir, battery, radiator, hydraulic systems, and brake fluid shall be checked and replenished as needed.
3. The fuel and air filters shall be changed.
4. The apparatus chassis and functional equipment shall be lubricated as per the manufacturer's specifications.
5. All moving rods and linkages shall be lubricated as required.
6. Nuts and bolts, including the lug nuts on wheels, shall be checked and tightened as needed.
7. Auxiliary generators shall be serviced.

B. Drivers are responsible for ensuring that the work is performed. The completed paperwork shall be forwarded by the officers to the maintenance and logistics officer. The paperwork shall include a list of all parts and supplies used in performing the work and the number of personnel hours involved.

C. Items in need of repair shall be recorded on the Equipment Service Request Form and the requests shall be forwarded to the maintenance and logistics officer.

D. Officers shall record the performance of any semiannual maintenance activity in their Daily Log.

VII. **Annual Maintenance**

The following items shall be performed on an annual basis:

A. The motor vehicle inspection (MVI) sticker shall be renewed during the appropriate month each year. Apparatus drivers shall be responsible for scheduling and renewing the MVI sticker.

B. Aerial devices shall undergo an annual service test in accordance with NFPA 1914, *Testing Fire Department Aerial Devices*. The test shall be performed by a qualified, independent testing firm.

1. The annual test shall include a visual inspection prior to an operational or load testing to note any visible defects, damage, or improperly secured parts.

2. An inspection of all the visible welds shall be made, as well as an inspection of all the bolts, cables, rollers, pins, slides, and washers. Bolts shall be torqued to the manufacturer's specifications.
3. A nondestructive test of the aerial device shall also be performed.
4. The test shall be scheduled and supervised by the maintenance and logistics officer.

C. Fire pumps:
1. A service test shall be conducted on all fire pumps at least once a year or whenever a pump has undergone extensive repair.
2. Service tests shall include a dry vacuum test, a priming test, a capacity test, a tachometer and engine rpm check, a relief valve test, an overload test, a 200-psi test, a 250-psi test, and a tank-to-pump flow test.
3. These tests will be scheduled and conducted by the maintenance and logistics officer. The results of the tests shall be recorded on the Annual Fire Pump Service Test Form.

D. Apparatus and motor vehicles:
1. The annual service shall be conducted on all apparatus and motor vehicles in accordance with the Preventive Maintenance Worksheet.
2. The maintenance and logistics officer shall schedule and supervise the performance of the maintenance.

VIII. Responsibilities

A. Officers are responsible for the care and maintenance of all motorized vehicles assigned to their command and shall adhere to all established maintenance schedules.

B. Drivers are responsible for the readiness of their assigned vehicles and shall perform all daily, weekly, monthly, quarterly, semiannual, and annual maintenance tasks as indicated on the maintenance and daily work schedules. Drivers are also responsible for the performance of scheduled maintenance on all reserve vehicles garaged at their station.

C. The maintenance and logistics officer is responsible for reviewing all checklists and repair requests to monitor the status of the fleet. The MLO shall schedule all repair work and shall be responsible for ensuring that the work has been satisfactorily performed. The MLO shall prepare a written report each month for the fire chief summarizing the maintenance activity performed during the previous month.

Maintenance, SOP 501.02

EQUIPMENT ASSIGNED TO AN APPARATUS

I. Scope

This standard governs the inventory of small tools and equipment carried on ambulances and fire apparatus. It was promulgated to:

A. Establish a minimum complement of small tools, equipment, and supplies to be carried on ambulances and fire apparatus operated by the department.

B. Establish a system to inventory, replenish, replace, discard, and repair small tools, equipment, and supplies carried on an ambulance or fire apparatus.

II. General

A. Ambulances and fire apparatus shall be assigned a specific complement of small tools, equipment, and supplies to be carried on each type of vehicle. A current, complete copy of the inventory assigned to each apparatus shall be carried on the apparatus at all times and shall be updated whenever an item is added to or deleted from the inventory.

B. Small tools, equipment, and supplies shall be carried in the same compartment or location on all apparatus of the same type whenever it is possible to do so. Unfortunately, the number and configuration of compartments will vary by the age, type, and manufacturer.

C. Ambulances and fire apparatus shall be inspected at the beginning of each shift and after every major incident to ensure that all small tools, equipment, and supplies are in their proper place, clean, and functioning properly.

D. Small tools, equipment, and supplies shall not be moved, added, deleted, or otherwise altered except on the approval of the fire chief.

E. Compartment designations have been standardized as follows:

L = Left side, driver's side, or port side.
R = Right side, officer's side, or starboard side.
T = Tailboard, backstep, rear, stern, or aft.
H = Hosebed.
B = Basket or compartment open to the top, usually located above the pump or cab.

F. Compartments are sequentially numbered from front to back, left to right, and top to bottom and will be designated by side.

III. Responsibilities

A. Officers are responsible for:

1. Marking and labeling all small tools, equipment, and supplies assigned to each piece of apparatus assigned to their command.
2. Reviewing the Driver's Daily Apparatus Checklist for accuracy and completeness.
3. Reporting the damage, destruction, or theft of any item.
4. Informing their relief of any changes that have taken place during their tour of duty.
5. Inspecting each apparatus under their command on a monthly basis to determine that all equipment is in its proper place, clean, and in proper working condition. The results of the quarterly inspection shall be included in the monthly report to the fire chief, and a copy shall be forwarded to the hose and equipment officer.

B. Drivers are responsible for:

1. Marking and labeling all items assigned to their vehicles.
2. Inventorying all small tools, equipment, and supplies assigned to an apparatus under their care at the beginning of their tour of duty and as otherwise required. Any addition, deletion, change, or loss shall be reported on the Driver's Daily Apparatus Checklist.
3. Informing their relief of any changes that have taken place during their tour of duty.

C. EMS program managers are responsible for:
   1. Maintaining an up-to-date inventory of all small tools, equipment, and supplies carried on all ambulances.
   2. Purchasing, replacing, and discarding small tools, equipment, and supplies assigned to ambulances and other EMS vehicles.
   3. Inspecting ambulances and other EMS vehicles on a monthly basis to ensure that the proper inventory is in place, drugs and supplies have not expired, and all items function properly. The results of the inspections shall be included in the manager's monthly report to the fire chief.

D. SCBA officers are responsible for:
   1. Maintaining an up-to-date inventory of all SCBA equipment assigned to apparatus, ambulances, and individuals.
   2. Purchasing, replacing, repairing, and discarding SCBA equipment assigned to apparatus and ambulances.
   3. Supervising the monthly SCBA inspection and reporting the results in his monthly report to the fire chief.

E. Equipment officers are responsible for:
   1. Maintaining an up-to-date inventory of all hose, ladders, and equipment assigned to ambulances and fire apparatus.
   2. Purchasing, replacing, repairing, and disposing of hose, ladders, and equipment assigned to apparatus and ambulances with the approval of the fire chief.
   3. Assigning and scheduling the annual testing of fire hose and ladders assigned to all apparatus and stations.

F. Haz mat officers are responsible for:
   1. Maintaining an up-to-date inventory of all reference materials, protective clothing, monitoring and detection equipment, chemicals and foams, equipment, and supplies assigned to ambulances and fire apparatus for use in a hazardous materials incident.
   2. Purchasing, replacing, repairing, and discarding supplies and equipment assigned to apparatus and medic units for use in a hazardous materials incident.
   3. Inspecting each ambulance and fire apparatus on a monthly basis to determine that all supplies and equipment are in their proper place, clean, and functioning properly.

G. Maps and street index officers are responsible for:
   1. Maintaining an up-to-date inventory of all maps and mapbooks owned by the department.
   2. Assigning mapbooks to individuals and to apparatus and updating the mapbooks as required. A master log of all updates shall be maintained.
   3. Inspecting all ambulances, fire apparatus, and support vehicles on a monthly basis to determine that all mapbooks are in their proper location, clean, and current.

Maintenance, SOP 501.03

DECLARING A VEHICLE UNSAFE TO OPERATE

I. Scope

This standard applies to ambulances, fire apparatus, and support vehicles owned or operated by the department. It was promulgated to:

A. Ensure that ambulances, fire apparatus, and other motorized vehicles are safe to operate by identifying a list of major defects that would render a vehicle unsafe to operate.

B. Establish a procedure to be used by the operator of a vehicle to place a vehicle judged to be unsafe out of service.

II. Procedure for Placing a Vehicle Out of Service

A. The operator of any apparatus or motorized vehicle shall have the authority to place the apparatus or vehicle out of service whenever one or more of the following defects are discovered:

1. A failure of the braking system that results in the vehicle being difficult or impossible to stop.
2. A failure of the windshield wipers during inclement weather. Note: On a sunny day, the failure of the wipers would not be a justifiable reason to remove the vehicle from service.
3. The failure of the headlights, brake lights, or taillights during a period of darkness.
4. A flat tire.
5. The inability to engage or operate a fire pump.
6. A failure of the hydraulic system or other component that prevents the operation of an aerial device.
7. A failure of the power steering system.
8. A failure of the coolant system causing the engine to overheat.
9. A failure of the HVAC system in the patient compartment of an ambulance during periods of temperature extremes where the failure to provide heat or cooling would compromise patient care.
10. Battery, alternator, or electrical system failure that prevents the vehicle from being started or that results in a discharge of the electrical system.
11. Failure of the mobile two-way radio if a portable is unavailable to replace the mobile radio.
12. Any defect that, if not immediately corrected, would cause further damage to the apparatus or vehicle or would endanger the lives of either the general public or the personnel assigned to it.

B. Whenever a defect is discovered in an apparatus or vehicle that routinely responds to emergency incidents that would require the apparatus or vehicle to be placed out of service, the officer or person responsible for the apparatus or vehicle shall notify Dispatch that the apparatus or vehicle is out of service.

C. If there is a reserve apparatus or vehicle available to replace the unit being placed out of service, the officer or person in charge of the apparatus or vehicle shall transfer personnel and all necessary equipment to the reserve and notify Dispatch of the change when the company is back in service.

D. If a reserve is unavailable, the fire chief or deputy chief shall be contacted and asked for instructions as to the disposition of personnel and equipment.

E. The officer or person in charge of the apparatus or vehicle shall be responsible for having the repairs made that will return the apparatus or vehicle to service.

   1. During normal business hours, the maintenance and logistics officer (MLO) shall be notified about the defects so that repair work can be scheduled.
   2. On nights, weekends, and holidays, the defects shall be noted on the Equipment Service Request Form. The completed form shall be forwarded to the MLO.

F. Whenever an apparatus or vehicle is placed out of service, a Red Tag Out-of-Service Card shall be taped to the inside of the windshield so that it obstructs the driver's normal range of vision.

III. Items That Require Repair but Do Not Justify Placing a Vehicle Out of Service

A. Any defect shall be repaired as soon as possible. Most defects do not necessitate placing an apparatus or vehicle out of service even though the defect may be annoying or might hamper normal operations.

B. Most notable would be the failure of one or more audio or visual warning devices. Since most apparatus have two sirens and an air horn, the failure of any one of these would not cause the vehicle to be placed out of service or prevent the unit from running emergency calls. The failure of all of the audio warning devices or warning lights would prevent the unit from running emergency calls, but would not require that the unit be placed out of service in the event that a reserve were not available.

IV. Responsibilities

A. It is everyone's responsibility to ensure that the vehicle or apparatus to which they have been assigned is safe to operate. Whenever any defect is discovered, it shall be reported in accordance with departmental policies and shall be repaired as soon as possible.

B. The burden of determining whether or not a vehicle is safe to operate shall rest primarily with the assigned driver of the vehicle. The driver has the primary responsibility of maintaining the vehicle and is the one crew member most familiar with its operational characteristics.

C. Officers shall be responsible for ensuring that vehicles assigned to their command are in proper working order and are properly maintained.

D. The MLO is responsible for reviewing all of the daily checklists and repair requests so as to monitor the overall status and condition of the fleet. The MLO shall also be responsible for confirming that all regularly scheduled maintenance is performed and coordinating all off-site maintenance and repair work.

Maintenance, SOP 501.04

MARKING VEHICLES AND APPARATUS

I. **Scope**

Apparatus and motor vehicles owned and operated by the department shall be properly marked and identified as required by state law and the applicable city, county, or district ordinances. (Vehicles operated by the fire marshal's office primarily for law enforcement purposes are exempt from the requirements of this standard.) This standard was promulgated to:

A. Establish a uniform system for marking and identifying apparatus and motorized vehicles.

B. Identify the station assignment of ambulances and fire apparatus.

C. Comply with the various local, state, and federal regulations requiring that vehicles owned by a municipality be clearly marked and identified as publicly owned vehicles.

II. **Marking Requirements**

A. All ambulances, fire apparatus, and other motorized vehicles owned or operated by the department shall be marked and identified with the department's logo decal. One decal shall be placed on each side of the vehicle and a decal shall be placed on the rear of the vehicle whenever practical.

B. Whenever possible, the decals will be placed on the front doors of the vehicle. The lettering on the decals shall be three-inch swirled gold leaf and shall be shaded in black. The decals shall be constructed of a reflective material and shall be stacked to read per the department's logo.

C. The lettering on ambulances shall consist of 10-inch red reflective letters when on a white background. Letters shall have a 1/4-inch black trim and shall be centered on each side of the module. The lettering shall be stacked to read per the department's MICU (Mobile Intensive Care Unit) logo.

1. Lettering on the rear of an ambulance shall display the name of the department and shall consist of four-inch red reflective letters with 1/4-inch black trim and shall be centered below the warning lights and above the rear entrance doors. The MICU in 10-inch white reflective letters with 1/4-inch black trim shall be centered on the rear entry doors.
2. The word *Ambulance* shall be placed on the front engine cowling of the ambulance in such a manner as to be read properly from the rearview mirror of a vehicle in front of the ambulance.
3. The name of the fire department in four-inch red reflective letters with 1/4-inch black trim shall be placed on the front of the patient compartment, centered near the top.
4. A 12-inch Blue Star of Life decal with 1/4-inch white reflective trim shall be centered on the chassis doors on each side and on the rear of the module.

III. **Painting Vehicles and Apparatus**

A. Ambulances and fire apparatus:

1. The main bodies of all ambulances and fire apparatus shall be painted red, as approved by the fire chief.

2. The tops of the vehicles shall be painted white, as approved by the fire chief.

B. Staff and support vehicles shall be painted fleet white.

C. Unmarked vehicles shall be painted as directed by the fire marshal.

IV. Striping and Numbering

A. A single four-inch-wide reflective stripe shall be placed on each side of a fire apparatus at or near where the headlights of an oncoming vehicle would be expected to strike the side of the apparatus. The stripes shall run the full length of the vehicle and be accompanied by gold stripes with black borders along the top and bottom edges of the white stripes.

B. On smaller apparatus, the stripes may be reduced to a three-inch width or may increase to larger sizes on large apparatus such as trucks and water tenders.

C. Staff and support vehicles shall be exempted from the striping and numbering requirements. However, the vehicle identification number will be placed on the left front fender.

D. Vehicle identification numbers shall be placed in a conspicuous location on all four sides of an apparatus. The numbers will normally be white and 12 inches in height. On ambulances and newer apparatus, the numbers are removable and interchangeable.

E. Vehicle identification numbers shall consist of at least two digits. The prefix will be an alphabetical character that identifies the classification of the apparatus. The second digit will be a numerical character that corresponds to the apparatus' station assignment.

V. Logos and Promotional Signs

A. 911 logos:

In an effort to promote the use of the 911 emergency telephone system to report emergencies, all marked vehicles shall have one or more 911 logos affixed to them. The logo shall be placed in a conspicuous location. The number of logos affixed to a vehicle as well as the size and color of the logos shall depend on the size and use of the vehicle.

B. Promotional materials or advertising:

No promotional materials or advertising shall be placed or displayed on any vehicle that belongs to the department unless the material or advertising is promoting a fire prevention or public safety program and has been approved by the fire chief.

VI. Restrictions

A. No bumper sticker, decal, logo, or sign, whether temporary or permanent, shall be affixed to or displayed on any vehicle or apparatus owned or operated by the department without the express written consent of the fire chief.

VII. Responsibilities

A. The fire chief shall establish and maintain a system that readily identifies that fire apparatus and motor vehicles operated by the department are the property of the department. The marking system shall also identify vehicles as to type and station assignment. Markings shall comply with all applicable NFPA standards.

B. The maintenance and logistics officer shall have the appropriate decals, lettering, numbers, and reflective striping affixed to vehicles as required by this standard.
C. Officers shall ensure that all vehicles assigned to their command be properly marked as required by the provisions of this standard.
D. Drivers shall be responsible for changing and updating the numbers on the apparatus that they are assigned to drive, provided that the vehicle is equipped with an interchangeable numbering system. The numbers shall correspond with the station and apparatus assignment.

Maintenance, SOP 502.01

MARKING AND INVENTORYING EQUIPMENT

I. **Scope**

This standard establishes an accountability system for the management of the material assets of the department, excluding buildings and motorized vehicles. It was promulgated to:

A. Prevent the loss or theft of material assets by providing an identification and labeling system.
B. Identify the station or apparatus that an item is assigned to.
C. Establish an inventory system for material assets.
D. Establish responsibilities for the acquisition and repair of equipment.

II. **Labeling**

A. All nonconsumable items shall be marked or labeled by some permanent means. The mark or label shall identify the item as being the property of the department.
B. Generally, items may be marked or labeled as follows:
   1. Cloth and other porous items shall be marked with indelible ink.
   2. Metallic items shall be marked with an electric engraver or die stamps.
   3. Apparel and protective clothing items shall be marked in accordance with the appropriate section of the Rules and Regulations Manual.
C. Bar coding:

   Whenever practical to do so, communications equipment, furniture, tools, and portable equipment shall have the department bar code label affixed, and the bar code number shall be recorded in the master inventory.
D. Color coding:

   Each station or major piece of apparatus will be assigned an exclusive primary color with which to mark its equipment.
   1. The color coding shall be done in a neat but conspicuous manner. On multiple company operations, the color coding will facilitate returning equipment to its assigned apparatus.
   2. All engine and truck companies shall be marked with their assigned primary colors.
   3. Ambulances will be marked with two colors. The first color will be the primary color of the engine company with which the ambulance is housed. The second color will be navy blue.

4. Brush units, tenders, and utility units will also use two colors. In addition to the primary color of the engine that the unit is housed with, brush units will use gold, tenders will use light green, and utility units will use purple.

E. Fire hose:

Each section of fire hose shall be stamped with the name of the department, its unique inventory number, and the month and year that it was tested for acceptance per SOP 503.02, Fire Hose Testing.

III. INSPECTION AND INVENTORY OF TOOLS AND EQUIPMENT

A. The department shall record and maintain a master list of all equipment, tools, furniture, and other material assets.

B. The inventory shall be updated each time a new piece of equipment is acquired or an old one is discarded.

C. The master inventory list should include the following:

1. Inventory number.
2. Description of the item.
3. Date the item was acquired.
4. Purchase price, vendor, and purchase order number.
5. Assigned location.

IV. Responsibilities

The following individuals shall be responsible for:

A. Communications officer:

1. Maintaining an up-to-date inventory of all two-way radio equipment, station alerting systems, repeaters, antennas and towers, base stations, pagers, and related accessories.
2. Purchasing, labeling, and issuing communications equipment to individuals, companies, and stations.
3. Supervising the repair of defective communications equipment.

B. EMS program manager:

1. Maintaining an up-to-date inventory of all medical supplies and equipment.
2. Purchasing, labeling, and issuing supplies and equipment to individuals, companies, ambulances, and stations.
3. Replacing and disposing of expired drugs and supplies.
4. Supervising the repair of defective EMS equipment.

C. Maintenance and logistics officer:

1. Maintaining an up-to-date inventory of all tools, spare parts, and related equipment.
2. Purchasing, labeling, and issuing tools and related equipment to companies, ambulances, and stations.
3. Supervising the repair of tools and related items.

D. SCBA officer:

1. Maintaining an up-to-date inventory of all SCBA and respiratory protection equipment.
2. Purchasing, labeling, and issuing equipment to individuals, companies, and stations.

3. Supervising the inspection, testing, and repair of SCBA and respiratory protection equipment.

E. Uniform and protective clothing officer:
   1. Maintaining an up-to-date inventory of all uniforms and protective clothing.
   2. Purchasing, labeling, and issuing uniforms and protective clothing to department personnel.
   3. Supervising the repair of uniforms and protective clothing.

F. Wellness officer:
   1. Maintaining an up-to-date inventory of all wellness and physical fitness equipment.
   2. Purchasing, labeling, and issuing equipment to fire stations to be used in wellness activities.
   3. Maintaining wellness equipment and supervising the repair of defective equipment.

G. Equipment officer:
   1. Maintaining an up-to-date inventory of all fire hose, ladders, and loose equipment owned by the department.
   2. Scheduling and supervising the required testing and repair of these items.
   3. Purchasing, labeling, and issuing these items to stations, companies, and individuals as appropriate.

H. Information management officer:
   1. Maintaining an up-to-date inventory of all computer hardware, software, and related accessories owned by the department.
   2. Purchasing, labeling, and issuing these items to stations, companies, and individuals as appropriate.
   3. Scheduling and supervising the maintenance of equipment and software.

I. Haz mat officer:
   1. Maintaining an up-to-date inventory of all equipment, disposables, chemicals and foams, protective clothing, reference materials, and related items used for hazardous materials incidents.
   2. Purchasing, labeling, and issuing these items.
   3. Supervising the repair of items used for hazardous materials incidents.

J. Maps and street index officer:
   1. Maintaining an up-to-date inventory of all maps used and issued by the department.
   2. Purchasing, labeling, and issuing maps according to departmental guidelines.
   3. Keeping all mapbooks, etc., up to date and supervising the replacement of out-of-date materials.

K. Officers:
   1. Marking and labeling all material assets assigned to their command.
   2. Reviewing the Driver's Daily Apparatus Checklist to ensure accuracy and completeness.

3. Reporting the damage, destruction, or theft of any item.

L. Drivers:

1. Marking and labeling all items assigned to their vehicles.
2. Conducting an inventory of the items assigned to their vehicles at the beginning of their tour of duty.
3. Recording any discrepancies on the Driver's Daily Apparatus Checklist.

Maintenance, SOP 503.01

FIRE HOSE

I. Scope

This standard establishes guidelines for the maintenance and storage of fire hose. It was promulgated to:

A. Establish guidelines for the proper care of fire hose.

B. Establish a procedure for promptly repairing damaged sections of fire hose.

C. Create and maintain an inventory of spare fire hose that can be placed in service whenever it is needed.

II. Cleaning

A. The life expectancy of a section of fire hose is determined by the care it receives. Hose is susceptible to mechanical injury, heat and fire damage, mold and mildew, and damage due to chemical contact and excessive pressures.

B. Each section of fire hose shall be inspected and cleaned after each use. All dirt, oil, and other foreign matter should be carefully removed by either clear water, a brush, or a mild soap or detergent. The hose should then be rinsed thoroughly.

C. After being cleaned, sections of fire hose should be properly dried unless the hose is of single-jacket, rubber construction. Rubber hose may be loaded wet after it has been cleaned. It is still possible, however, for mold to develop if wet hose is stored for a long period of time without proper drying. This may result in damage to the hose.

III. Repair

A. Whenever a section of hose is taken out of service and sent to the hose shop for repair, it shall be cleaned, dried, and properly red-tagged. The tag shall include a description of the defect.

B. All female hose couplings need to be fitted with gaskets. Couplings should be inspected periodically, and gaskets that have deteriorated or are missing should be replaced. A stock of spare gaskets is maintained in the workshop at each fire station for this purpose.

C. Petroleum products should not be used on threads or any part of a coupling.

D. To prevent permanent damage, the hose on each apparatus shall be reloaded every six months unless the entire load has been laid at a fire or for training. During one of the reloads, the annual hose test will be conducted.

E. As a general rule, no vehicle should drive over a hose. During an incident, however, it may become necessary for an emergency vehicle to drive over a hoseline. In such a case, the line should be charged to reduce the likelihood of damage.

IV. **Storage**

A. While the bulk of the spare hose inventory shall be stored at a single location, each station shall be assigned an inventory of spare hose of varying sizes as storage space and the master inventory permit.

B. After the hose has been properly cleaned and dried, the spare sections shall be rolled for storage. The hose should be rolled with the male coupling inside, then placed on a hose rack until it is needed.

C. Hose racks shall be placed in well-ventilated areas and out of contact with direct sunlight.

D. Hose shall not be stored out of doors or left on drying racks.

V. **Responsibilities**

A. The hose and equipment officer shall ensure that each section of fire hose is permanently marked and numbered as prescribed by SOP 502.01, Marking and Inventorying Equipment.

B. Officers shall maintain a complete record of all fire hose assigned to their company or station. The record shall include the date a section of fire hose was received, test records, in-house repairs, and the date a section was sent to the hose shop for repairs. This information shall also be included as a part of the company log.

C. Drivers shall be responsible for maintaining the proper quantity of fire hose on their apparatus. Hose loads shall be maintained in a neat manner, and all hose shall be loaded in accordance with SOP 603.04, Hose Loads.

Maintenance, SOP 503.02

FIRE HOSE TESTING

I. **Scope**

This standard regulates the acceptance of new sections of fire hose and establishes procedures for conducting the annual service tests. It was promulgated to provide a reasonable degree of assurance that the fire hose, couplings, and nozzles used by the department will perform as designed.

II. **Test Procedure**

A. Prior to testing, each section of hose shall be subjected to a physical inspection to determine whether it is free of debris; exhibits any evidence of mildew or rot; or is damaged from chemicals, burns, cuts, abrasion, and vermin. Any section of hose that fails the physical inspection shall immediately be placed out of service and sent to the hose shop for repair.

B. Hose shall be tested by using the pump of a reserve engine. The test area shall be relatively flat and free of any objects that might damage the hose.

C. The service test for hose of less than five inches in diameter shall be conducted as follows:

1. Connect the hose to an engine. Hose shall *not* be attached to any discharge at or adjacent to the pump operator's position.
2. The total length of any hoseline in the test layout shall not exceed 300 feet. Hoselines shall be straight and without kinks or twists. Hose that has been repaired or recoupled shall be tested one section at a time.

3. Connect the engine to a hydrant.
4. Connect a nozzle or shutoff device to the end of the hose. The appliance should be secured to prevent an uncontrolled reaction in the event of a hose rupture.
5. Fill the hoseline to be tested with water and bleed off all trapped air.
6. Close the nozzle and increase the pressure to 50 psi. Check for leakage. Tighten couplings as necessary. Mark the location of the couplings with a suitable marker.
7. Clear the area and increase the pressure slowly until the pressure reaches 250 psi for a service test or 400 psi for an acceptance test if manufactured prior to July 1987. Hose manufactured after July 1987 shall be tested to the pressure marked on the hose jacket. Hold for five minutes. Inspect for leaks or damage. Remember: Never straddle a hoseline! Consult NFPA 1962, *Standard for the Care, Use, and Service Testing of Fire Hose Including Couplings and Nozzles,* if you have any questions about this matter.
8. Bleed off pressure on conclusion of the test.
9. Mark all hose that passes the test with the month and year.
10. Record the test date, etc., in the permanent hose record.
11. Hose that fails the test by bursting or leaking or because of coupling failure due to slippage or leakage shall be tagged and placed out of service.
12. After the test, all hose shall be cleaned, drained, and dried before being placed in service or storage.

D. Tests for five-inch supply line and sections of soft suction hose shall follow the same procedure outlined in Item C, above, except that the service test pressure shall be 200 psi and the acceptance test pressure shall be 400 psi. Ensure that the hose is service tested while lying flat.

E. Booster hose shall be tested to 110 percent of its maximum working pressure.

F. Hard suction hose shall be tested on an annual basis using a dry-vacuum test of 22 inches hg for 10 minutes. If used under positive pressure, the test shall be repeated using 165 psi of water pressure.

G. Nozzles and other appliances shall also be inspected during the annual fire hose service test to ensure that the nozzles and appliances are undamaged, clear of obstructions, and fully operational. Any nozzle or appliance found to be in disrepair shall be red-tagged, removed from service, and sent for repair.

III. Responsibilities

A. The hose officer shall be responsible for:

1. Ensuring that all new sections of fire hose purchased by the department are designed and constructed in accordance with the provisions of NFPA 1961, *Standard on Fire Hose.*
2. Conduct an acceptance test on each section of fire hose before it is placed in the hose inventory. The test shall comply with the provisions of NFPA 1962, *Standard for the Care, Use, and Service Testing of Fire Hose Including Couplings and Nozzles.*
3. Schedule and supervise the annual service test. Every section of fire

hose in the department's inventory shall be tested in accordance with NFPA 1962.

4. Conduct a service test after a section of hose has been repaired.

B. Officers shall be responsible for all fire hose assigned to their command. Any section of fire hose that is discovered to be defective, is improperly marked, or has an out-of-date test shall be taken out of service. The hose officer shall be contacted for instructions on correcting the situation.

Maintenance, SOP 503.03

FIRE HOSE RECORDS

I. **Scope**

This standard establishes an inventory and record-keeping system for fire hose. It was promulgated to:

A. Provide an inventory of all fire hose owned by the department.

B. Provide a history of the service life of each section of fire hose.

C. Determine the amount of hose carried on an apparatus.

D. Determine the location and status of each section of fire hose at any given time.

II. **Master Hose Record**

A. The hose officer shall maintain a separate record for each section of fire hose. The record shall be maintained as long as the section of hose remains in the hose inventory. Hose records shall be included in the master inventory of the department's physical assets.

B. Fire hose records shall be maintained on the Master Hose Record Card.

C. The top three lines of the card shall be completed by the hose officer on the acceptance of a section of hose.

D. The remaining portion of the card shall be used to record all subsequent tests and repairs.

III. **Hose Book**

A. Each apparatus assigned a complement of fire hose shall carry a hose book. The hose book shall be used to maintain an up-to-date and accurate inventory of all the fire hose carried on the apparatus.

B. Any changes in the fire hose inventory carried on a fire apparatus shall be recorded in the hose book, along with the date that the changes occurred, the reason for the changes, and the name of the supervisor who made the changes. Changes might include adding, removing, or relocating one or more sections of hose on the apparatus.

C. The identification number of a section of fire hose shall be recorded in the hose book at the time the section is loaded onto an apparatus. The date that the section of hose was placed on the apparatus shall also be recorded.

D. Whenever hose is loaded on an apparatus to replace sections that were removed, the numbers of any sections that remain in the bed from the previous load shall be transferred to the next column. The numbers of the new sections that are added to complete the load shall then be recorded.

Maintenance, SOP 504.01

SELF-CONTAINED BREATHING APPARATUS

I. **Scope**

This standard establishes guidelines for the inspection and maintenance of self-contained breathing apparatus (SCBA). It was promulgated to:

A. Provide a reasonable degree of assurance that an in-service SCBA will function properly.

B. Require that any SCBAs that do not function properly be removed from service and repaired.

C. Comply with the applicable rules, regulations, and standards concerning SCBA equipment.

II. **General**

A. All SCBA equipment shall comply with the provisions of the edition of NFPA 1981, *Standard on Open-Circuit Self-Contained Breathing Apparatus for the Fire Service,* that was in effect at the time that the equipment was purchased.

B. SCBA equipment shall also comply with the applicable rules, regulations, and standards promulgated by other appropriate agencies.

III. **Inspection**

SCBA equipment shall be inspected periodically to determine its readiness for use and to discover and repair any damage or excessive wear sustained by the unit. The frequency of inspection is as follows:

A. Before and after each use:

1. Before each use, each SCBA assigned to an apparatus should be inspected for the following:

   a. Cylinder pressure: The minimum pressure should not fall below 90 percent of the full cylinder pressure. Example: A 4,500-psi cylinder should not contain less than 4,050 psi.

   b. Low-air alarm: This should sound when the cylinder valve is opened.

   c. Facepiece: This should be clean and free of debris, and it should seal properly to the wearer's face.

   d. Exhalation valve, bypass value, and operation: All should function normally.

2. After each use, each SCBA shall be inspected for the following:

   a. Low or empty cylinder: Cylinders should be cleaned, low cylinders should be refilled, and empty cylinders should be replaced with fully charged ones.

   b. Components and facepieces: Shall be cleaned, checked for excessive wear or damage, sanitized, and checked for proper function.

   c. Caution: Anytime an SCBA has been used in a contaminated atmosphere, the unit shall be completely decontaminated prior to being returned to service.

B. Daily inspection:

1. Each piece of SCBA equipment assigned to an apparatus shall be inspected at the beginning of each shift and after every use.

2. The inspection shall be made to ensure that the SCBA is fully charged, clean, free of damage, and fully operational.
3. Cleaning, refilling cylinders, or minor repair shall be done by the driver of the apparatus as needed. (Some jurisdictions require that all repairs be performed by a certified technician.)
4. Any unit showing damage or that does not function properly shall be removed from service and red-tagged with a description of the defect. The defects shall be noted on the Driver's Daily Apparatus Checklist.

C. Monthly inspection:
1. Each SCBA shall be inspected monthly. The inspection shall consist of those items listed on the Monthly SCBA Inspection Form.
2. The monthly inspection shall be performed on the first day of each month. Any unit that fails the inspection and that cannot immediately be repaired shall be red-tagged with a description of the defect, and the unit shall be placed out of service.
3. All spare cylinders and SCBA units shall also be inspected.
4. The completed checklists shall be forwarded to the SCBA officer.

D. Semiannual inspection:
1. All SCBAs shall be inspected and serviced every six months by the SCBA officer or by an authorized repair facility.
2. The inspection and servicing shall include the following:
    a. Disassembly and cleaning of the regulator and other major components, such as the low-air alarm, facepiece, etc.
    b. Replacement of all worn parts.
    c. Reassembly of the SCBA and testing for proper function.

E. Annual inspection:

Internal inspections of SCBA cylinders shall be performed annually to determine any condition that may contribute to the deterioration of a cylinder. Checks shall be made for rust, corrosion, moisture, damage, and evidence of oil or hydrocarbon contamination.

IV. **Breathing Air Supplies**

A. All breathing air produced for use in an SCBA shall comply with the testing and quality requirements of the Compressed Gas Association G-7.1 Commodity Specifications for Air, for Grade D Air. Tests shall be conducted quarterly by an independent testing service to determine whether the air quality meets or exceeds these requirements. The test results shall be filed and posted.

B. The breathing air that is produced by the department's compressors meets the air quality standards for SCBA equipment. It does *not* meet the requirements for SCUBA equipment. Therefore, no member of the department should ever fill a SCUBA tank from department sources.

C. Any air cylinder suspected of containing contaminated air or air that does not meet the department's air quality standards should be emptied and purged.

D. Any air cylinder that is not used within a period of three months shall be emptied and refilled. This shall be done on the assigned quarterly motor day for each company.

E. All SCBA cylinders shall be maintained at 90 percent of their rated storage capacity as stamped on the cylinder. Cylinders containing less than 90 percent of their rated capacity shall be segregated from full cylinders until they are refilled.

F. The SCBA officer shall ensure that all fire suppression personnel are capable of properly operating all breathing air compressors and cascade systems. Proper operating procedures and precautions shall be posted in a conspicuous location at each fill station.

G. Prior to filling a cylinder, personnel shall visually inspect all cylinders to ensure that none are damaged or defective and that they are within the current hydrostatic test date. Defective or out-of-date cylinders shall not be filled and shall be removed from service. (NFPA 1500, *Standard on Fire Department Occupational Safety and Health Program,* 5-3.7.3 requires that cylinders be hydrostatically tested within the applicable periods specified by the manufacturer and the applicable government agency. For example, if a manufacturer required a cylinder to be tested every five years and the last test date on the cylinder was 5/91 and the inspection took place on 9/97, the cylinder would be out of date and would have to be tested prior to being filled.)

H. All cylinders shall be refilled in accordance with the recommendations of the manufacturer.

I. Breathing air compressors shall not be operated in a contaminated atmosphere. Prior to operation, personnel shall ensure that the intake area is free of obvious contaminants. Air shall not be taken while any apparatus motor is running near an intake area.

V. Records

A. The SCBA officer shall ensure that the following records are kept:

1. A complete inventory of all SCBA units, cylinders, cascades, fill stations, special tools, spare parts, and related equipment.
2. Individual records for each regulator and harness assembly. The records shall include inventory and serial numbers, date of purchase, vendor, manufacturer, P.O. number, assigned location, maintenance and repair history, history of parts replacement, upgrades, and performance tests.
3. Individual records shall be kept for each cylinder. The records shall include inventory or serial numbers, date of purchase, vendor, manufacturer, P.O. number, assigned location, hydrostatic test dates, and a history of inspection and repair.
4. Individual maintenance and repair records shall be kept for each breathing air compressor, cascade system, fill station, purification system, and any ancillary equipment used to produce or store breathing air.

B. The SCBA officer shall also keep a permanent file of all quarterly air quality tests.

VI. Responsibilities

A. The fire chief shall appoint an officer to the collateral duty of SCBA officer.

B. The SCBA officer shall be responsible for the management and oversight of the SCBA inspection and maintenance program, including all cascade systems, breathing air compressors, and the mobile cascade system carried on the rehab unit. The SCBA officer shall ensure the timely repair of any defective piece of SCBA equipment.

C. Officers shall be responsible for the care and maintenance of the SCBAs, cascade systems, and breathing air compressors assigned to their command.

D. Every person assigned to fire suppression duty shall be responsible for knowing how to use and care for the SCBAs assigned to his apparatus. In addition, each firefighter shall be issued his own individual SCBA facepiece. The facepiece shall be kept clean and in proper working condition.

E. It is the duty of each individual to promptly report and correct any deficiencies found with any piece of SCBA equipment. If the equipment cannot immediately be repaired, the equipment shall be taken out of service.

Maintenance, SOP 505.01

GROUND LADDERS

I. Scope

This standard establishes guidelines for the maintenance, inspection, and service testing of ground ladders.

II. General

A. Ground ladders used for firefighting purposes shall be constructed in compliance with the provisions of NFPA 1931, *Standard on Design of and Design Verification Tests for Fire Department Ground Ladders.*

B. Ground ladders used for nonfirefighting purposes shall be constructed in compliance with the applicable OSHA standards.

III. Inspection and Care

A. Ground ladders shall be inspected and tested as required by NFPA 1932, *Standard on Use, Maintenance, and Service Testing of Fire Department Ground Ladders.*

B. A visual inspection shall be performed on the 15th day of each month and after every use. The inspection shall include, but not be limited to, the following items:

1. Check the heat sensors, if present, for evidence of heat exposure.
2. Test the rungs for snugness and tightness.
3. Check the bolts and rivets for tightness.
4. Check the welds for cracks and defects.
5. Check the beams and rungs for cracks, splintering, breaks, gouges, checks, wavy conditions, or deformation.
6. Check the butt spurs for excessive wear or other defects.
7. Conduct an operational check of the roof hook assemblies on roof ladders.
8. Check the halyards and wire cables on extension ladders for snugness, wear, and defects.
9. Clean, lubricate, and conduct an operational check of the pawl assemblies on extension ladders.
10. Lubricate the ladder slide areas as needed.

C. Aluminum and fiberglass ladders shall be waxed with an automotive paste wax as needed to maintain the finish, to inhibit corrosion, and to inhibit surface deterioration.

D. Wooden ladders shall be protected by at least two coats of a clear spar varnish. If the coating becomes damaged, the area should be sanded and new varnish applied. Ladders shall be refinished as needed.

E. Ground ladders shall be maintained as free of moisture as possible and shall be wiped after being sprayed with water or used in the rain.

F. Ground ladders shall not be painted except for the top and bottom 18 inches of each section for purposes of identification or visibility. When painted, the top sections shall be painted white and the bottoms black.

G. Any ladder that shows signs of failure during the visual inspection shall be removed from service and either repaired or destroyed. A wooden ground ladder shall be removed from service and tested if dark streaks develop in the beams.

IV. <u>Annual Service Test</u>

A. Ground ladders shall be service tested annually in accordance with the provisions of NFPA 1932. The tests shall be conducted during June of each year. In addition, a service test shall be conducted whenever a ladder is suspected of being unsafe; has been subjected to overloading, heat exposure, or direct flame contact; or has other unusual conditions and after any repair work.

B. Use caution when performing service tests on ground ladders to prevent damage to the ladder or injury to personnel during testing. The test load shall be placed on the ladder in a manner so as to avoid any shocks or impact loading. Personnel involved in the testing should be aware of the potential for sudden and dramatic failure of the ground ladder undergoing the service test.

C. Any sign of failure during the service test shall be sufficient cause for the ladder being removed from service and repaired or destroyed as ordered by the fire chief.

D. The results of a service test shall be recorded on the Ground Ladder Record Form.

E. The tests to be performed are:

1. Horizontal bending test, except for folding ladders.
2. Roof hook test for roof ladders.
3. Hardware test for extension ladders.
4. Hardness service test for metal ground ladders.
5. Folding ladder horizontal bending test.
6. Rung-twist test.

F. Service tests shall be scheduled by and conducted under the supervision of the maintenance and logistics officer.

Section 600

Emergency Operations

Emergency Operations, SOP 600.01

GLOSSARY OF TERMS

I. Scope

This standard lists and defines common terms used by operations personnel to manage fire control and rescue incidents.

II. Definitions

1. Aid station: Where triage takes place and necessary EMS personnel and equipment are located to provide aid to fire victims and fire service personnel. The person in charge may be either a fire service or an EMS person. The person assigned should be at least paramedic certified. All activities should be coordinated through the command post.
2. All clear: The primary search has been completed.
3. Command post: The location at which the primary command functions are executed. The command post is manned by the incident commander and other support personnel as required. The command post shall designate frequencies to be used by command support operations.
4. Incident commander (IC): The person with overall responsibility for a particular incident. The person will use the radio term *Command* and may use a geographical identifier when multiple operations occur—e.g., "Main Street Command."
5. Rural water supply—terminology
   A. Attack unit: Usually the first piece of fire apparatus on the scene, the attack unit sets up hoselines to fight the fire. It can consist of more than one unit if necessary.
   B. Drafting: Using a pump to lift water from a supply source, such as a river, pond, ditch, dump tank, etc.
   C. Fill site: The location where the tank trucks go to get loads of water, such as a hydrant, draft site, storage tank, etc.
   D. Rural fire: More than 1,000 feet from a fire hydrant.
   E. Tanker (or tank truck): A fire truck used primarily to carry large quantities of water for rural firefighting. Also known as a water tender.
   F. Tanker shuttle: Using several tank trucks to transport water from a water supply source to a fire scene.
   G. Working fire: A fire that will require considerable effort to extinguish and may require an additional response of apparatus, such as water tenders in rural operations.

6. Sector: A smaller, more manageable unit of command delegated by the incident commander to provide management and command for specific functions or geographical areas. Sectors shall be designated as Sector 1, 2, 3, etc., and shall be assigned on a clockwise basis. For example:

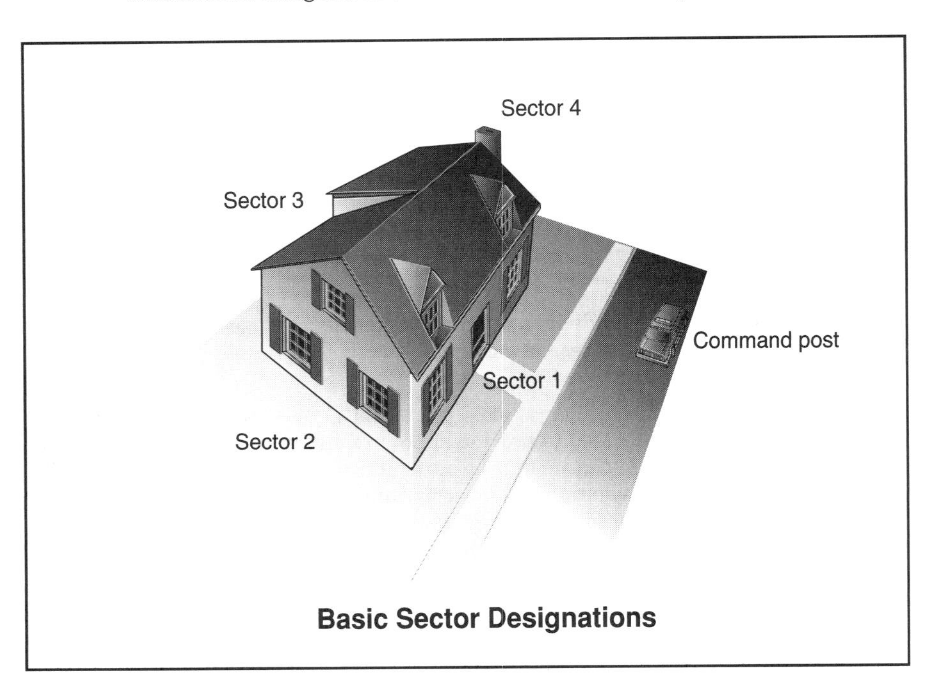

Basic Sector Designations

For high-rise operations, the sector above the fire shall be designated before the sector below and be numbered according to the floor number. For example, if a fire is on the 21st floor of a 30-story building, the sector above the fire would be Sector 22. Other sectors may be designated by location or function, such as roof, interior, rehab, triage, transportation, etc.

7. Sector commander: A person in charge at a given area of the fireground. The sector commander shall be assigned by the incident commander and coordinate operations through the command post and shall operate on the command frequency.
8. Staging area: That location where incident personnel, apparatus, and equipment are assigned in an available status. The staging officer shall coordinate activities through the command post. If necessary, the command post shall assign a staging frequency that will normally be the primary alarm channel. Dispatch shall be responsible for move-ups and shall report them to the staging officer.
9. Support officer: An aide assigned to the incident commander and normally located in the command post. The incident commander may require all activities such as water supply, staging, etc., to be coordinated through the support officer.

Emergency Operations, SOP 600.02

INCIDENT COMMAND SYSTEM

I. <u>Scope</u>

This standard establishes guidelines for the management of fire and rescue incidents.

II. <u>General</u>

A. It shall be the policy of this department to implement the incident command system (ICS) for all fires, haz-mat incidents, rescues, and EMS responses that require two or more ambulances.

B. The ICS shall also be used at all incidents in which the scope and complexity of the incident exceeds the capabilities of routine operations.

III. <u>Command</u>

A. There are three levels or types of command:

1. Forward command
2. Mobile command
3. Fixed command

B. <u>Forward command</u>: Means to lead from the front and is the exact opposite of a fixed command position. Forward command is task-oriented and should be used:

1. When the incident requires further investigation.
2. When the incident requires an immediate, aggressive, hands-on action and it may not be appropriate to pass command.
3. During any incident by company commanders who may not be in charge of a specific sector or function but who are actually performing an assignment—e.g., manning an interior hoseline, ventilation, search and rescue, etc.

C. <u>Mobile command</u>: The commander moves around. Normally this should be limited to sector commanders.

D. <u>Fixed command</u>: A command post is established in a safe, secure environment that allows for efficient, effective operations.

E. <u>Passing command</u>: Command is passed from one incident commander to another by the following means:

1. The first-arriving officer may designate another officer as Command if it is known that the other officer will arrive momentarily.
2. Command may be passed at the command post on arrival of a senior, more experienced command officer at the request of Command or on the senior officer's assessment that it is necessary to assume command.

IV. <u>Establishing Command</u>

A. On initial arrival, the first-arriving officer or unit shall:

1. Notify Dispatch by radio of his arrival.
2. Provide a brief size-up of the situation found on arrival. The following information shall be provided at minimum:
   a. <u>Fires</u>: (1) The size and construction of the building or situation encountered and (2) the extent of involvement.
   b. <u>EMS</u>: "Out checking" or "One-car rollover," etc.

3. Instruction to other units.
   a. Designation of Command.
   b. Level I or Level II staging.
   c. Request for additional assistance—e.g., a second alarm.
   d. Orders—e.g., lay supply line, reduce code, disregard, etc.
   e. Actions you are taking—e.g., investigation, pulling preconnect, etc.

B. Subsequent arrival of other units:
   1. Notify Dispatch of arrival.
   2. Notification of status—e.g., "Engine Two staged at the hydrant at Elm and McKinney."
   3. Ask Command for instructions.

C. Incident commander:
   1. Shall use the term *Command* on all radio traffic. May also use a geographical designation—e.g., "Elm Street Command."
   2. Shall identify the type of command (i.e., mobile or fixed) and the location of the command post.
   3. Shall designate sectors.
      (a) Fires: Staging, Rehab, Sector 3 (rear), etc.
      (b) EMS: Transportation, Rehab, Triage, Staging, etc.
   4. Designate support functions: Safety, Water Supply, etc.
   5. Shall complete Tactical Worksheets on all incidents in which the system is used.
   6. Shall terminate command whenever appropriate.
   7. Shall conduct a postfire critique or postmortem.

Emergency Operations, SOP 600.03

MINIMUM COMPANY STANDARDS

I. Scope

This standard establishes minimum standards of performance for fire and rescue companies. It was promulgated to maintain the proficiency of fire and EMS companies by establishing a program for the annual evaluation of company-level skills.

II. General

A. This standard establishes a series of evaluations for fire and rescue operations. These evaluations simulate tasks commonly performed during routine incidents. Each fire and rescue company is required to be able to perform all of these evolutions within the allotted time frame.

B. Each fire and rescue company shall be evaluated in December of each year to ensure that the company can meet the prescribed minimum standards.

III. Responsibilities

A. Company officer: Each company officer shall train with his respective company on at least a weekly basis. A company officer is responsible for the performance levels of his assigned company, and his company shall be capable of successfully performing all of the evolutions contained within this standard.

B. Deputy fire chief: The deputy fire chief shall evaluate each company annually on these minimum company standards and shall forward the results of the evaluations to the fire chief. The deputy fire chief shall assist any company whose performance is below standard and shall schedule companies for reevaluation when necessary.

C. Shift training officer: The STO shall assist companies on their respective shifts in maintaining proficiency in basic fire and rescue skills and will provide training aids and materials available as well as technical assistance.

D. Fire chief: The fire chief shall require all companies to be evaluated on an annual basis in December and shall assign an officer to schedule and conduct the evaluations.

IV. Evaluation Procedures

A. Company evaluations are based on performance and time standards.

1. The department has established a maximum time limit to properly complete each evolution.
2. Performance is measured by observing the correct procedures and techniques established by the department.
3. Safety in all phases of an evolution is a prime consideration and shall not be compromised for speed.
4. If possible, the annual evaluation will take place at a site selected to simulate realistic fireground conditions, such as abandoned buildings or buildings under construction.
5. During the annual evaluations, the evolutions will simulate actual emergency operations. This means that company members will wear full protective clothing and SCBA.

B. Companies shall be scheduled annually and will be evaluated on two of the fire evolutions, one EMS evolution, and one equipment operator standard. The results shall be recorded on the Minimum Company Standards Evaluation Form.

C. There are established time and performance standards for each evolution. Performance errors (task errors) will be assessed as additional time and will be added to the actual time taken to complete the evolution. A company whose total time (actual time plus errors) exceeds the time standard will be scheduled for a reevaluation of that evolution. The evaluator may fail any company that commits serious safety violations or whose performance is unsatisfactory.

D. Companies that do not successfully complete an evolution will be retested within 30 days. If the company fails the retest, appropriate disciplinary action will be taken.

E. After the company has completed an assigned evolution, it may be assigned additional tasks from the list of basic tasks described below. An error or safety violation that occurs when basic tasks are being performed will be noted on the evaluation form and may justify additional training and reevaluation. Fine adjustments to engine pressure, relief valve, and governor setting, as well as removing kinks, must be done to avoid task errors. However, this will not be a part of the timed portion.

BASIC TASKS

Evolution: EMS 1—MAST APPLICATION

Time Standard: Five minutes.

Personnel Required: One ALS ambulance crew, one engine or truck company (two to four personnel).

Procedure:

1. Unfold MAST (Medical Anti-Shock Trousers) and place on full backboard (the over-the-arm method may also be used).
2. Apply trousers:
   A. Legs.
   B. Abdomen—to bottom of ribcage.
3. Secure self-fasteners or zippers.
4. Connect pump.
5. Open stopcock to legs.
6. Close stopcock to abdomen.
7. Inflate legs (continually checking blood pressure).
8. Close stopcocks to legs.
9. Open stopcock to abdomen.
10. Inflate abdomen (if indicated).

Evolution: EMS 2—TRACTION SPLINT APPLICATION

Time Standard: Five minutes.

Personnel Required: Minimum of two.

Procedure:

1. Check motor, sensory, and circulatory functions.
2. Measure splint.
3. Position straps.
4. Apply traction hitch to ankle.
5. Apply manual traction and support extremity.
6. Position traction device and attach ishial strap.
7. Tighten winch to obtain traction.
8. Secure leg to splint.
9. Check security of extremity to splint.
10. Check motor, sensory, and circulatory functions.
11. Secure patient to long board.
12. Repeat motor and sensory check in extremities.

Evolution: EMS 3—KED™ APPLICATION

Time Standard: Eight minutes.

Personnel Required: One ALS ambulance crew, one engine or truck company (two to four personnel).

Procedure:

1. Perform motor and sensory check in all extremities.

2. Apply and maintain cervical traction.
3. Apply cervical collar.
4. Position short board or Kendrick Extrication Device™ (KED™) behind patient.
5. Secure patient to short board. Fill voids.
6. Secure patient head to short board.
7. Evaluate immobilization.
8. Lift patient onto long board.
9. Lower patient's legs to long board.
10. Secure patient to long board. Fill voids.
11. Repeat motor and sensory check in extremities.

Evolution: EMS 4—TRAUMA ASSESSMENT

Time Standard: Ten minutes.

Personnel Required: Minimum of two personnel.

Procedure:

1. Establish level of consciousness.
2. Establish airway (consider possible C-spine injury).
3. Check breathing rate.
4. Check for chest trauma, assess breathing quality, treat, administer O_2.
5. Check circulation rate and quality.
6. Identify and correct any severe bleeding.
7. Check neck and apply appropriate stabilization (including chest).
8. Check vitals (blood pressure only if radial pulse is present).
9. Complete secondary survey.
10. Check vitals.

Evolution: EMS 5—THUMPER APPLICATION

Time Standard: Eight minutes.

Personnel Required: One ALS ambulance crew and/or one engine or truck company (two to four personnel).

Procedure:

1. Place thumper-adapted backboard next to patient on patient's left side.
2. Connect thumper control board to the backboard.
3. Place patient on the backboard and secure patient with shoulder straps provided.
4. Connect oxygen supply to the control board at master valve (No. 1).
5. Connect thumper compression arm column assembly in the control panel, position it over the lower half of the sternum, and set in stroke limited position.
6. Open master valve (No. 1) only after checking to ensure force control valve (No. 3) is off.
7. Open cardiac compression valve (No. 2).
8. Adjust force control (No. 3) by turning clockwise until proper compression depth is obtained.
9. Open ventilation valve (No. 4) to start ventilation using mask provided.
10. Adjust oxygen supply valve (No. 5) to proper delivery setting.

Evolution: FIRE 1—USE OF SCBA

Time Standard: One minute.

Personnel Required: Single company (three to four personnel).

Procedure:

1. Don SCBA from the normal position on the apparatus. (The captain and driver may step out of the apparatus to don SCBA.)
2. Time will start when the evaluator says "Go."
3. Time will stop when the basics have been completed. This means that:
   A. The SCBA is operating in the positive-pressure mode.
   B. The waist belt is fastened.
   C. The facepiece is properly sealed.
   D. The protective hood is on correctly with no skin exposed.
4. Failure to properly complete any of these four steps will result in failure of this evaluation.
5. Note: Gloves do not have to be worn. This is a company standard, and *all* members must complete the test within the time allotted. The failure of any one individual will result in the failure of the entire company.
6. This evolution does *not* replace the individual quarterly drill requirements.

Evolution: FIRE 2—LADDER RAISE

Time Standard: 14-foot ladder: Thirty seconds.
24-foot ladder: One and one-half minutes.
35-foot ladder: Two minutes.

Personnel Required: Single company (three to four personnel).

Note: This evolution will be performed with full personal protective equipment and SCBA. The SCBA does not have to be in service and the facepiece does not have to be worn.

1. 14-FOOT LADDER STAND:

   One firefighter removes and raises the 14-foot ladder. Time will start when the evaluator says "Go." Time will stop when the ladder is ready to climb.
2. 24-FOOT LADDER/TWO-PERSON STAND/FLAT OR BEAM RAISE:

   Two firefighters will raise the 24-foot ladder to a second-story window or an equivalent height. Time will start when the person footing the ladder says "Go." Time will stop when the ladder is extended and secured. The evaluator will determine whether the flat or the beam raise is to be used.
3. 35-FOOT LADDER/THREE-PERSON STAND:

   Three firefighters will raise the 35-foot ladder to a third-story window or an equivalent height. Time will start when the captain gives the first command. Time will stop when the ladder is extended and the bangor knot is tied.

Evolution: FIRE 3—SETTING LIGHTS AND EXHAUST FANS

Time Standard: Three minutes.

Personnel Required: Single company (three to four personnel).

Procedure:

1. Time starts when personnel leave their seats and time ends when the evolution is completed.

2. Full protective clothing and SCBA will be worn. Personnel entering the building will have their SCBA and PASS device operating.
3. Start the generator.
4. Advance a portable floodlight to the second floor and place into service.
5. Position the floodlights on the apparatus and place them in service to illuminate the front of the building.
6. Place an exhaust fan (gasoline or electric) at the doorway, start the fan, and blow air into the building using the correct PPV technique.

Evolution: FIRE 4—TRUCK COMPANY OPERATIONS

Time Standard: Five minutes.

Personnel Required: One truck company (three to four personnel).

Procedure:

1. The truck will be positioned a short distance from the drill tower. When personnel are ready, the signal to begin will be given.
2. The driver of the truck will spot the truck in front of the tower. Time begins when the truck stops.
3. Steps of operation:
   A. The truck crew will raise a 24-foot ladder to the second floor and advance a chain saw, three axes, a pike pole, a 3/8-inch × 50-foot utility line, and a 14-foot roof ladder to the roof mockup.
   B. The captain and two crew members will then open a 4-foot × 4-foot hole in the roof.
   C. The fourth crew member, if available, will heel the extension ladder and assist on the ground.
   D. The crew will then descend the ladder.
4. Time stops when all crew members have descended the ladder.

Evolution: FIRE 5—SUPPLYING TWO 1¾-INCH HANDLINES

Time Standard: Four minutes for a four-member crew.
Four and one-half minutes for a three-member crew.

Personnel Required: One engine company (three to four personnel).

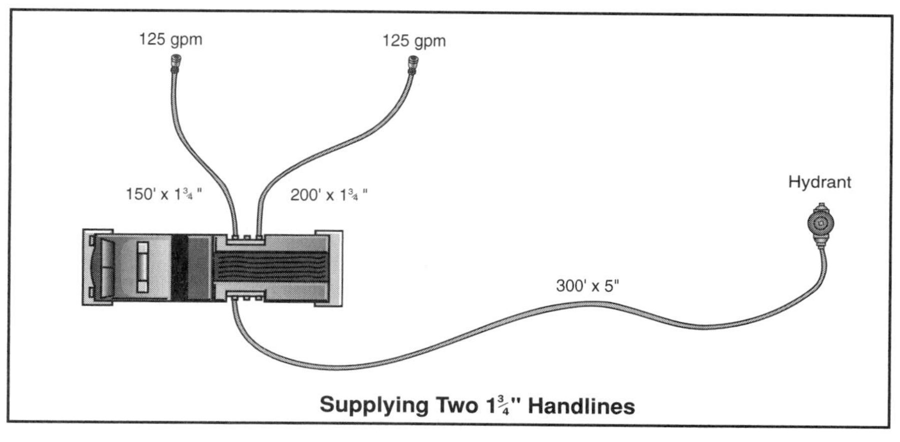

Supplying Two 1¾" Handlines

Procedure:

1. Stage the engine away from the hydrant. When personnel are ready, give the signal for the engine to proceed to the hydrant.
2. Time starts when the engine stops at the hydrant.
3. Steps of operation:
   A. Forward lay a single 5-inch supply line from the hydrant a distance of 300 feet.
   B. Advance two 1¾-inch preconnects (one 150 feet and one 200 feet).
   C. Flow 250 gpm at a nozzle pressure of 100 psi.
4. Time stops when both lines are supplied properly.

Evolution: FIRE 6—TWO-PIECE OPERATIONS

Time Standard: Five and one-half minutes for eight personnel.
Six minutes for six personnel.

Personnel Required: Two engine companies (six to eight personnel).

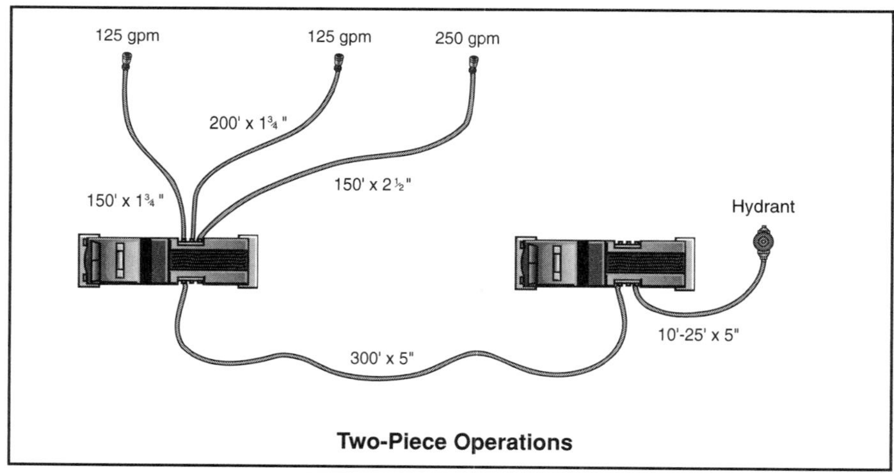

Two-Piece Operations

Procedure:

1. Stage both engines away from the hydrant. When personnel are ready, give the signal for the attack engine to proceed to the fire area.
2. Start the time when the attack engine stops at the fire.
3. Delay the second engine by 30 seconds.
4. Steps of operation:
   A. The attack engine stops at the fire and advances two 1¾-inch preconnects (one 150 feet and one 200 feet) and flows 200 gpm.
   B. The supply engine arrives from the opposite direction, delaying by 30 seconds. The supply engine stops, dropping off a supply line and its crew. The crew assists in hooking up the supply line, advances the 2½-inch preconnect, and flows 250 gpm.
   C. The driver of the supply engine then lays 300 feet of LDH (5-inch) supply line to the hydrant and relay pumps to the attack engine.
   D. Stop time when all lines are at the proper pressure and flow.

Evolution: FIRE 7—DECK GUN OPERATIONS

Time Standard: Three minutes.

Personnel Required: One engine company (three to four personnel).

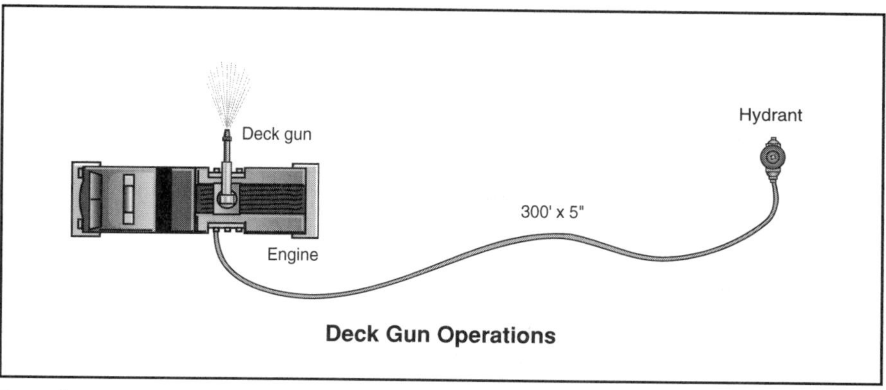

Deck Gun Operations

Procedure:

1. Stage or ready engine and personnel away from the hydrant.
2. When personnel are ready, give signal for engine to proceed to hydrant.
3. Time starts when engine stops at the hydrant.
4. Steps of operation:
   A. Lay one 5-inch supply line from the hydrant a distance of 300 feet.
   B. Place deck gun in operation and operate at proper pressure and flow.
5. Time stops when the deck gun is properly supplied.

Evolution: FIRE 8—ELEVATED MASTER STREAM

Time Standard: Five minutes.

Personnel Required: One engine company and one truck company (six to eight personnel).

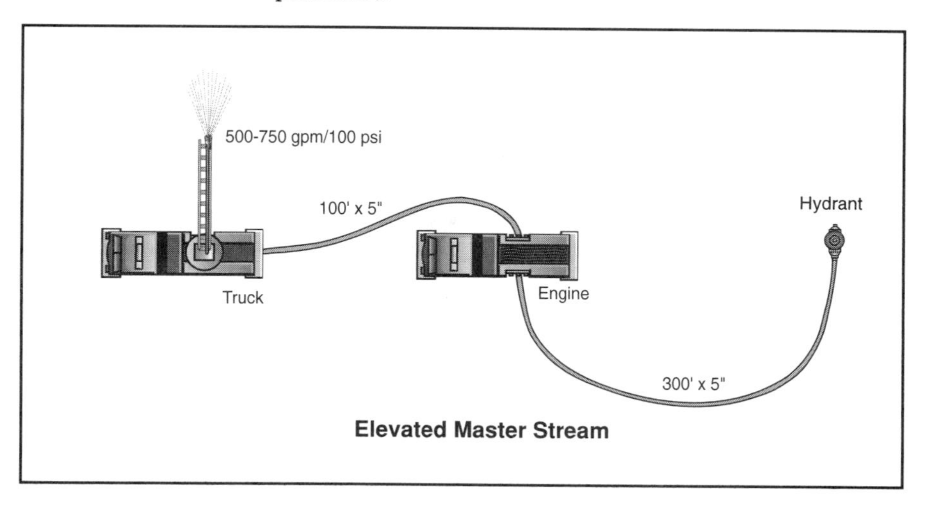

Elevated Master Stream

Procedure:

1. Stage or ready the engine, truck, and personnel away from the hydrant. When personnel are ready, give signal for engine to proceed to hydrant and truck to proceed to fire area.
2. Time starts when the engine stops at the hydrant.
3. Steps of operation:
   A. Lay one 5-inch supply line from hydrant a distance of 300 feet.
   B. Position the truck, raise the aerial device, and prepare ladder pipe for service.
   C. Lay one 100-foot supply line from engine to intake.
   D. Operate ladder pipe at proper pressure and flow.
4. Time stops when the ladder pipe is properly supplied.

Evolution: FIRE 9—SUPPORT OF AN AUTOMATIC SPRINKLER SYSTEM

Time Standard: Four minutes.
Personnel Required: One engine company (three to four personnel).

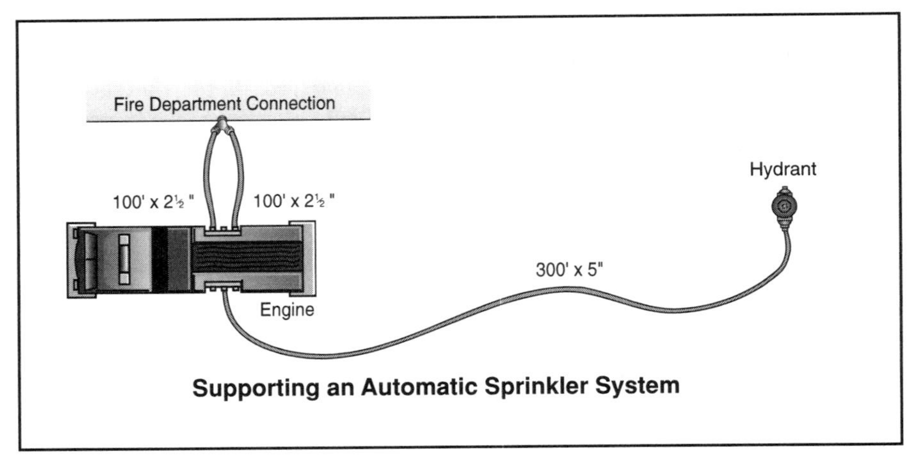

Supporting an Automatic Sprinkler System

Procedure:

1. Stage or ready engine and personnel away from the hydrant. When personnel are ready, signal the engine to proceed to the hydrant.
2. Time starts when the engine stops at the hydrant.
3. Steps of operation:
   A. Lay one 5-inch supply line from the hydrant a distance of 300 feet.
   B. Lay two 2½-inch supply lines a distance of 100 feet from the engine to the sprinkler connection.
   C. Operate all lines at proper pressures and flows (150 psi at fire department connection).
4. Time stops when all lines are properly supplied.

Evolution: FIRE 10—SUPPLYING A PORTABLE MASTER STREAM

Time Standard: Four and one-half minutes with four personnel.
Five minutes with three personnel.

Personnel Required: One engine company (three to four personnel).

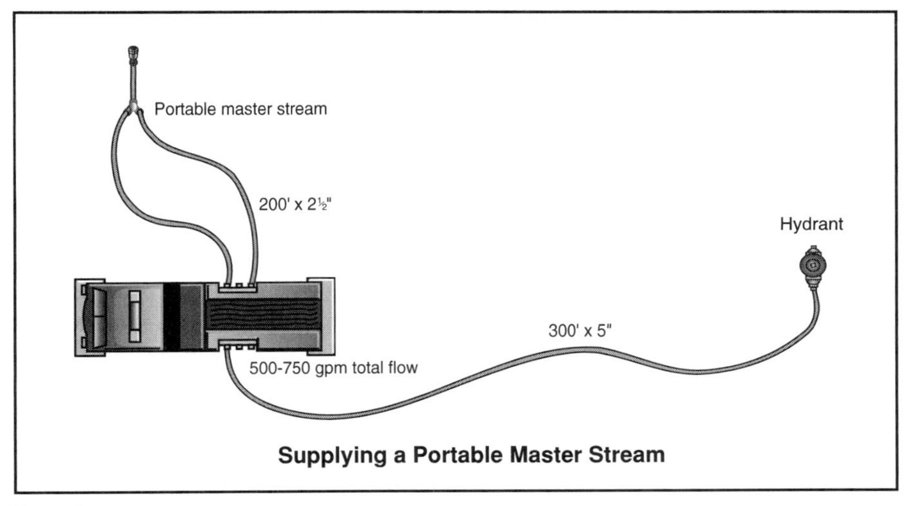

Supplying a Portable Master Stream

Procedure:

1. Stage or ready engine and personnel away from the fire area. When personnel are ready, signal the engine to proceed to the fire area.
2. Time starts when the engine stops at the hydrant.
3. Steps of operation:
   A. Forward lay one 5-inch supply line a distance of 300 feet from the hydrant.
   B. Advance two 2½-inch supply lines approximately 200 feet from the engine, deploy a portable master stream device, and connect the supply lines to the device.
   C. Supply 500 to 750 gpm to device using either a smooth-bore or fog nozzle.
4. Time stops when the proper flow rates are achieved.

Evolution: FIRE 11—KNOTS AND HITCHES

Time Standard: None. Pass or fail.

Personnel Required: Each member shall demonstrate competence.

Procedure:

1. All members assigned to respond to emergency incidents shall be able to demonstrate the ability to tie a minimum of seven standard knots and hitches.
2. Required knots and hitches:
   A. Bowline.
   B. Bowline on a bight.
   C. Clove hitch.
   D. Figure eight.
   E. Half hitch.

F. Square knot.
G. Timber hitch.

3. Personnel assigned to emergency operations should be able to tie and use the following knots and hitches:
   A. Becket or sheet bend.
   B. Life basket.
   C. Running bowline.

Evolution: DRIVER 1—SERPENTINE EXERCISE

Time Standard: Not applicable.

Equipment Required: One engine, truck, or ambulance plus three traffic cones.

Procedure:

1. Drive the apparatus in a straight line along the left sides of the traffic cones, stopping just beyond the last cone.
2. Back the apparatus between the cones by passing to the left of No. 1, to the right of No. 2, and to the left of No. 3. The driver shall not back the apparatus without someone assisting the driver as per department SOPs.
3. Stop the vehicle and drive it forward between the cones by passing to the right of No. 3, to the left of No. 2, and to the right of No. 1.
4. Running over or moving any of the traffic cones will cause the driver to fail the exercise.

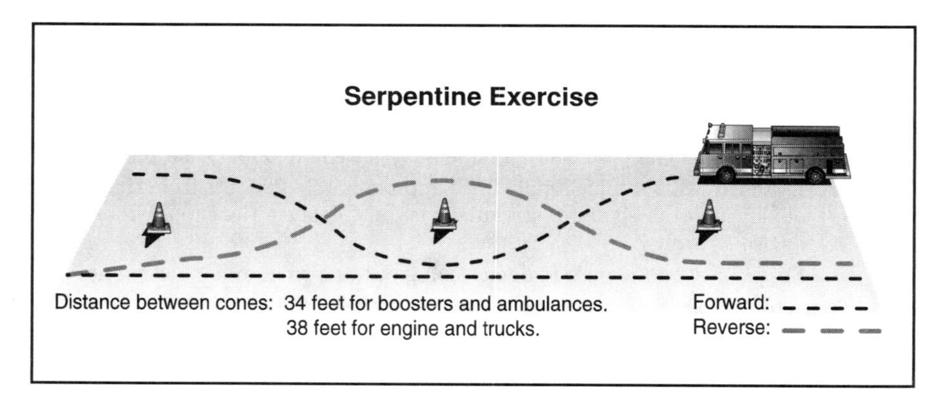

Evolution: DRIVER 2—ALLEY DOCK EXERCISE

Time Standard: Not applicable.

Equipment Required: One engine, truck, or ambulance plus traffic cones.

Procedure:

1. Drive past the barricades with the dock to left of the operator.
2. Back into the stall by making a left turn. The driver shall not back up the apparatus without someone assisting him as per department SOPs.
3. Running over or moving any traffic cones will cause the driver to fail the exercise.

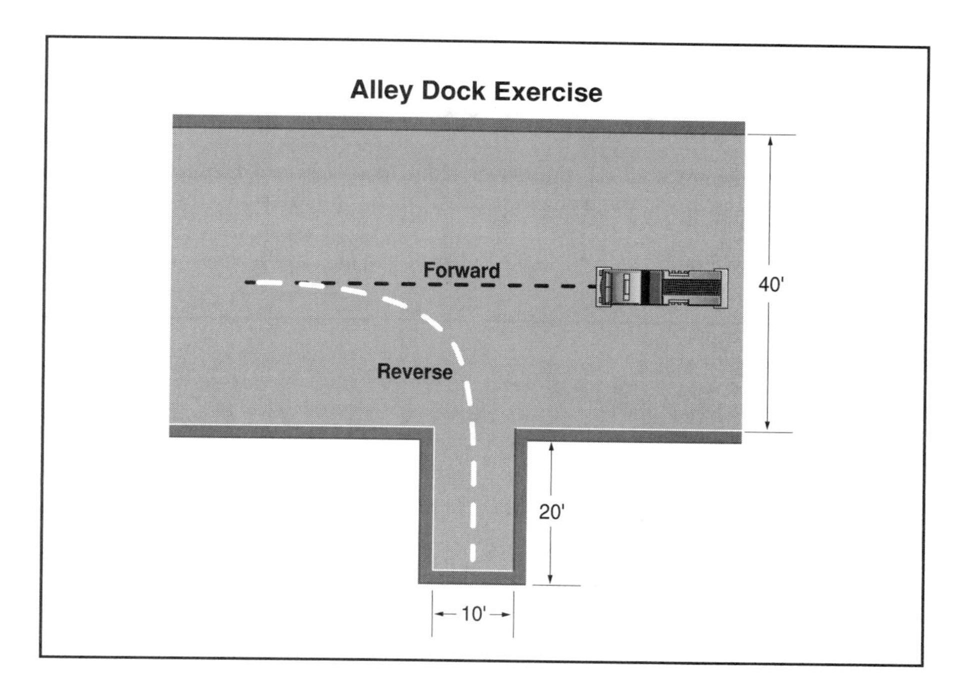

Evolution: DRIVER 3—OPPOSITE ALLEY EXERCISE

Time Standard: Not applicable.

Equipment Required: One engine, truck, or ambulance and traffic cones.

Procedure:

1. Traveling at a safe, moderate speed, the driver will move forward and steer the apparatus in a straight line into the opposite alley as indicated by the diagram below.
2. The driver shall not stop or back up the apparatus during the maneuver. Hitting or moving a traffic cone will cause the driver to fail the exercise.

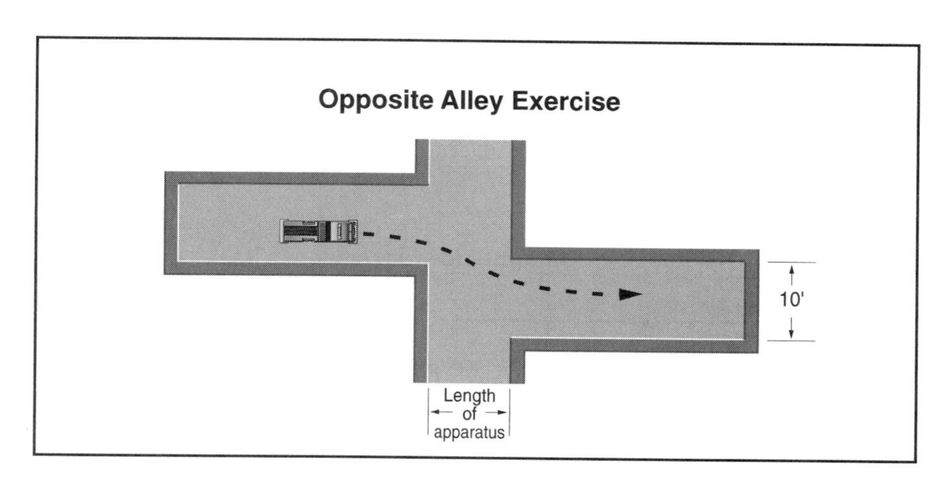

Evolution: DRIVER 4—DIMINISHING CLEARANCE EXERCISE

Time Standard: Not applicable.

Equipment Required: One engine, truck, or ambulance and traffic cones.

Procedure:

1. At a safe, moderate speed, the driver shall move forward through the course without hitting any of the traffic cones. Hitting a cone will cause the driver to fail the exercise.
2. The driver will stop the vehicle 50 feet past the last traffic cone with the front bumper on the finish line.

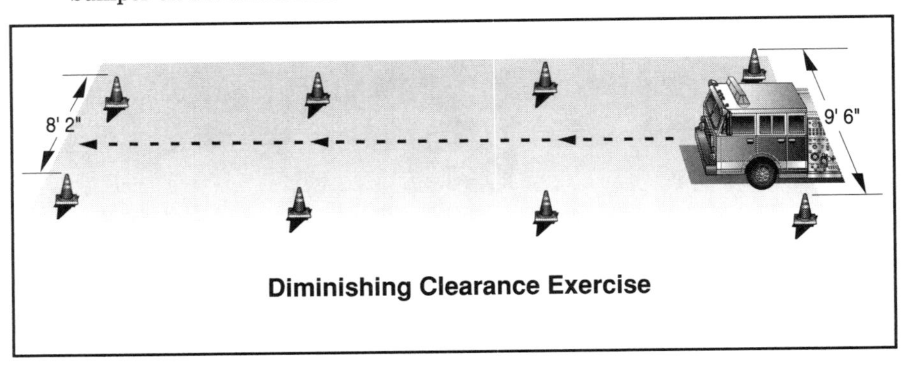

Diminishing Clearance Exercise

Evolution: DRIVER 5—STRAIGHT-LINE EXERCISE

Time Standard: Not applicable.

Equipment Required: One engine, truck, or ambulance and traffic cones.

Procedure:

1. The driver shall proceed through the course as indicated in the diagram below without stopping.
2. The driver shall accelerate through the gears.
3. Running over a traffic cone shall cause the driver to fail the exercise.

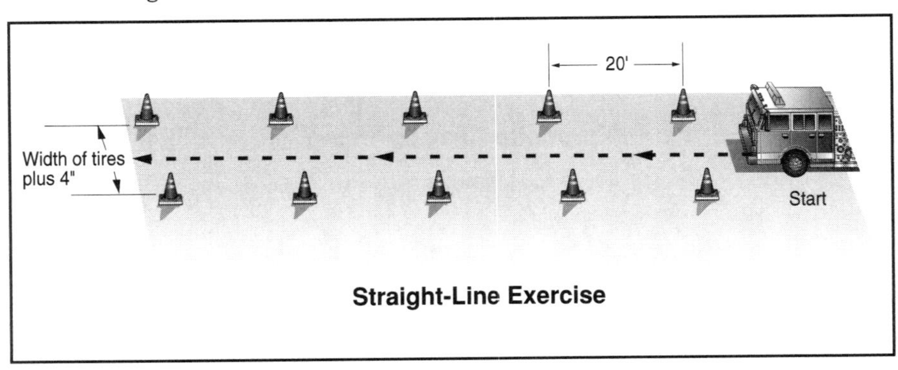

Straight-Line Exercise

Evolution: DRIVER 6—TURNING-AROUND EXERCISE

Time Standard: Not applicable.

Equipment Required: One engine, truck, or ambulance and traffic cones.

Procedure:

1. The driver shall move the vehicle in the opposite direction of original travel as per the diagrams below.
2. The driver shall not back up the apparatus without assistance. (Note: Department SOPs require someone to assist a driver when backing.)
3. Hitting or moving a traffic cone shall cause the driver to fail the exercise.

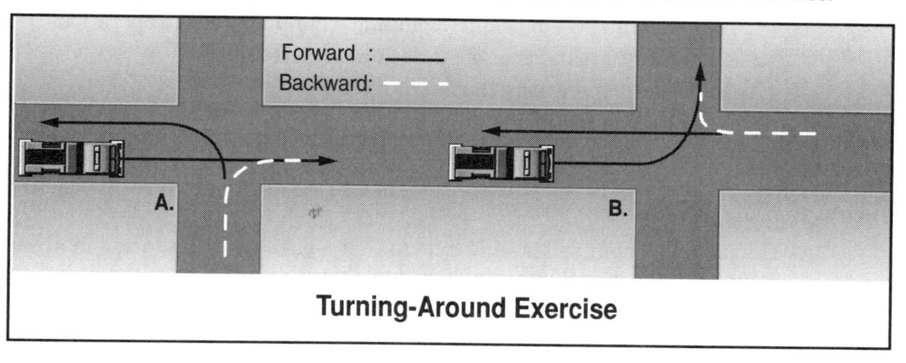

Turning-Around Exercise

Emergency Operations, SOP 600.04

POSTMORTEMS

I. Scope

This standard requires the completion of a written postmortem for all emergency incidents that involve one or more of the following:

1. A fire fatality.
2. More resources than the initial assignment
3. A mass-casualty incident.

This standard was promulgated to establish the guidelines for preparing and conducting a postmortem critique of a fire control or rescue incident.

II. Definitions

A. Critique: A postincident review of the problems encountered, what went right, and the lessons learned during a fire control or rescue incident.

B. Postmortem: A written report published after the review of a major incident that includes all the relevant facts concerning the incident. The report shall include a review of the problems encountered, the lessons learned, a summary of the on-scene operations, and any recommendations for changes in SOPs, etc.

III. Procedure

A. A postincident critique shall be conducted in the following circumstances:

1. A working structure fire—a fire that requires a considerable effort to extinguish and may require the response of additional apparatus such as water tenders in rural efforts.
2. A ground-cover fire that requires the response of three or more companies.
3. A haz-mat incident that requires a minimum of a first-alarm assignment.
4. An EMS incident that requires the response of three or more companies.

B. A critique shall be conducted as soon as practical following the incident and shall include:
   1. A review with the companies that responded to the incident.
   2. If practical, a review with the entire on-duty shift.
   3. A review of applicable SOPs.

C. In addition to a critique, a postmortem shall include:
   1. A narrative of the incident, including a summary of the operations.
   2. A review of problems or obstacles encountered.
   3. A review of operations that went well.
   4. A diagram of the incident.
   5. Recommendations for changes in SOPs, commendations, etc.
   6. A critique with all shifts.

IV. <u>Responsibilities</u>

A. The incident commander shall ensure that a critique is conducted in accordance with the guidelines established by this standard. The critique shall be conducted as soon as practical following the incident.

B. The incident commander shall also have a postmortem published and critiques conducted in accordance with this standard.

C. The incident commander shall also forward to the fire chief within 72 hours of an incident a narrative of the operations conducted, including statements by each company officer.

D. The fire marshal shall forward to the fire chief within 72 hours the following:
   1. A completed incident report, including fire alarm records.
   2. A summary of the initial investigation.
   3. Copies of all injury and causalty reports.

Emergency Operations, SOP 600.05

TACTICAL GUIDELINES

I. <u>Scope</u>

This standard shall regulate the management of all emergency incidents to which the department responds. It was promulgated to:

A. Establish rules and procedures to manage fire control and rescue activities.

B. Produce standard and predictable fire control and rescue results.

II. <u>Goals and Objectives</u>

A. A fire is the direct result of a failure of our fire prevention efforts. The best fires are the ones that never get started. Therefore, it is the goal of this department to prevent fires and to save lives and property.

B. In pursuit of this goal, it shall be the policy of the department to:
   1. Do no harm! Take no action that will cause further injury to a person or intentionally damage property.
   2. Be safe! Always obey all the safety rules and procedures promulgated by the department.
   3. Be nice! Always treat each other and members of the public with respect.

III. Operational Priorities

Three priorities *must* be addressed at every incident to which the department responds. These priorities are discussed below in order of importance.

A. Life safety/rescue:

1. Primary search: It shall be standard procedure to extend a primary search in *all* involved and exposed occupancies that can be entered safely. The completion of the primary search process is reported using the standard radio reporting term "*All clear.*"
2. Rescue efforts: Extend rescue efforts in the following order:
   a. The most severely threatened.
   b. The largest number of people.
   c. The remainder of the fire area.
   d. The exposed areas.
3. Firefighter safety: Due to the hazardous nature of firefighting, the safety of firefighters is of primary importance. Therefore, the following safety rules are to be observed:
   a. All persons involved in firefighting or other hazardous situations shall wear *full* protective clothing and SCBA. No one, regardless of rank, shall enter a hazardous atmosphere without SCBA.
   b. The incident commander and sector commanders shall be responsible for persons operating in hazardous locations and shall make sure that all personnel are accounted for.
4. Victim recovery: The fire department shall conduct a secondary search after the fire is out and shall assist in the recovery of all fire victims. On recovery, the appropriate authorities will be notified, and at no time shall the names of injured or deceased firefighters be mentioned on the radio.
5. Aid station: At large working fires, multiple alarms, etc., the rehab section will respond to the scene and set up an aid station in an accessible, uncongested area. The incident commander shall give whatever support is necessary to assist the aid station. The rehab unit will be released from the scene on order of the incident commander.
   a. At every incident, at least one paramedic will be assigned to monitor the physical condition of the members present. The paramedic shall have the authority to relieve members of their duties for a period of time, if necessary.
   b. The aid station will be responsible for reporting the status of the members under their supervision to the incident commander.
6. Safety officer: At least one member will be assigned as safety officer by the incident commander at a working incident. The safety officer shall have the authority to correct any violations of established safety SOPs. Following the incident, all violations will be reported to the incident commander in writing.

B. Fire control:

1. It is standard procedure to attempt to stabilize fire conditions by extending, wherever possible, an aggressive, well-placed, and adequate interior (offensive) fire attack effort and to support that aggressive attack with whatever resource and action may be required to reduce extension and to bring the fire under control.

2. Initial attack efforts must be directed toward supporting primary search and rescue operations.
3. Fire streams are to be operated only on fires, not into smoke.
4. The following operations are to be initiated at every incident:
   a. Size-up.
   b. Rescue/life safety.
   c. Exposure protection.
   d. Confinement.
   e. Extinguishment.
   f. Property conservation.
5. Write off property that is lost and protect exposed property based on the most dangerous direction of spread. Always attack structure fires from the *unburned* side! Do not continue operations in positions that are essentially lost.
6. <u>Fire stream management</u>:
   a. It is the responsibility of each engine company to provide its own uninterrupted, adequate supply of water. "Provide" does not necessarily mean to lay a supply line or pump the water, but rather to get an adequate, reliable supply of water into the pump by whatever means available.
   b. When in doubt, lay hose. The company officer shall make this decision. Remember, it is better to pick up a dry line that wasn't used than to need a line that wasn't laid out.
   c. Factors relating to the type of line pulled:
      (1) Size.
      (2) Placement.
      (3) Speed.
      (4) Mobility.
      (5) Supply.
   d. Booster lines shall not be pulled as the first line on working structure or automobile fires. Hoselines of 1½ inches shall be the minimum size pulled.
   e. Hoseline placement:
      (1) Place the first stream between the fire and the persons endangered by it.
      (2) If no life is endangered, place the first stream between the fire and the most severe exposure.
      (3) The second line should be taken to the secondary means of egress.
      (4) A third line should back up the first.
      (5) Assist rescue.
      (6) Protect exposures.
      (7) Support confinement.
   f. Operate heavy streams, if necessary, but *not* when an interior attack is taking place. Do not combine interior and exterior attacks. Before heavy exterior streams are operated, the incident commander shall instruct Dispatch to advise all personnel via radio.

g. Shut nozzles down when necessary. Do *not* operate into ventilation holes! Also, do not apply water to the roof in a mistaken effort to extinguish fire.

C. Property conservation:
   1. After rescue and fire control considerations, it shall be standard procedure to commit whatever fireground resources are required to reduce loss to an absolute minimum. All members are expected to perform in a manner that consistently reduces loss during fire operations.
   2. Property conservation activities shall include but are not limited to:
      a. Prompt interior and exterior fireground lighting.
      b. Proper ventilation.
         (1) Mechanical (PPV).
         (2) Natural.
      c. Salvage.
      d. Overhaul.
      e. Proper fire stream management.
   3. If necessary, the incident commander shall call for fresh or additional personnel to complete property conservation activities.

Emergency Operations, SOP 600.06

TACTICAL SURVEYS

I. Scope

This standard requires company commanders to prepare tactical surveys for target hazards within their company's first-due area. It was promulgated to:

A. Establish guidelines for preparing a standardized tactical survey of target hazards within the district.

B. Provide a system for making tactical surveys readily available to incident commanders.

II. Definitions

The following definitions will assist in the preparation and use of tactical surveys:

A. Construction type: NFPA 220, *Standard on Types of Building Construction,* identifies five types of construction:
   1. Type I: Fire-resistive construction.
   2. Type II: Noncombustible construction.
   3. Type III: Ordinary construction.
   4. Type IV: Heavy timber.
   5. Type V: Wood frame.

B. Fire flow: The rate of water flow needed to fight or confine a fire. Fire flows range from a minimum of 500 gpm for a single-family detached dwelling to a maximum of 12,000 gpm for large manufacturing or industrial complexes.

C. Fire load: The sum, measured in pounds per square foot, of the combustible contents and construction materials contained in a building. Fire loads vary from 5 lbs./sq. ft. in residential occupancies to more than 60 lbs./sq. ft. in hazardous occupancies.

D. Flame spread: A numerical expression of the relative rate at which flame will spread over the surface of a material. It is not an indication of fire resistance.
   1. Class A: 0–25.
   2. Class B: 26–75.
   3. Class C: 76–200.
   4. Class D: 201–500.
   5. Class E: Greater than 500.

E. Occupancy: An indication of the general use of a building. There are nine standard occupancy classifications:
   1. Assembly.
   2. Commercial.
   3. Educational.
   4. Hazardous.
   5. Industrial.
   6. Institutional.
   7. Mercantile.
   8. Residential.
   9. Storage.

F. Occupant load: The number of people present at the time of the emergency. Occupant loads vary by occupancy type and the time of day. Consider occupants with special needs when preparing the tactical survey—e.g., nonambulatory patients in a nursing home.

G. Risk: The probability of a fire or other emergency governed by existing conditions favoring the event and the presence of causative agents. Also known as a fire hazard.

H. Target hazard: A building or occupancy that is potentially hazardous to multiple loss of life or that could potentially require a significant commitment of resources to manage a fire control or rescue event.

I. Tactical survey: A written plan developed by the first-due fire company to manage a fire control or rescue event in a specific building or occupancy. The tactical survey is prepared in accordance with the recommendations contained in this standard.

III. Preparation of the Plan

A. According to NFPA 1420, *Recommended Practice for Pre-Incident Planning for Warehouse Occupancies,* a preincident plan is one of the most valuable tools available to aid the fire department in effectively controlling emergencies.

B. NFPA 1420 states that a preincident plan involves information gathering; analysis and dissemination; the "what-if approach"; and review, drill, and evaluation.

C. A preincident plan will be identified as a tactical survey.

D. Individual target hazards will be identified within each station's first-due area based on the possibility that a fire or other emergency could result in:
   1. Multiple life loss or injuries.
   2. A significant number of people placed in potential jeopardy.
   3. The required fire flow exceeding the capability of the first-alarm assignment.

E. The fire chief will assign each company officer with a list of target hazards on which to prepare tactical surveys.

F. The company officer will conduct a field survey on each target hazard to assist in the preparation of the tactical survey. Inspection files, building plans, plans for automatic fire suppression and detection systems, etc., may also prove useful.

G. The survey team should record the following information:

1. Construction type and building size.
2. Occupancy and occupant load.
3. Phone contacts and owner.
4. Site plan, accessibility, and any unusual security measures.
5. Fire load and contents.
6. Calculation of the fire flow requirement using the ISO formula:

   $F = 18\ C\ (A)^{0.5}$

   F = flow in gpm.

   C = coefficient related to construction type.

   A = total area of all floors except basement.

   The survey team should also identify the adequacy of the water supply and estimate the resources necessary to manage the event.
7. The presence or absence of built-in fire detection and suppression systems and an organized fire brigade, etc. Note the location of fire department connections, values, shutoffs, fire doors, etc.
8. Record the location of building services, utility shutoffs, meters, etc.
9. Note any unique features or special hazards that might present problems for responding personnel.

H. The survey results will be recorded on a Tactical Survey Form. The survey will be forwarded to the fire marshal for review. The fire marshal will forward the survey along with any comments to the fire chief for his final review and approval.

I. Copies of the approved tactical survey will be placed in the occupancy's inspection file. They will also be compiled into a manual to be kept on each apparatus and command vehicle.

IV. Responsibilities

A. The shift training officer shall schedule periodic drills to test and evaluate the information contained in the tactical surveys.

1. These tests should range from a single company tabletop to a full-scale functional exercise involving multiple companies.
2. The information gathered from these exercises should be used to evaluate the effectiveness of the tactical survey and to make corrections as needed.

B. Company officers shall be responsible for the preparation and maintenance of tactical surveys for target hazards within their respective first-due areas.

C. Drivers shall be responsible for ensuring that the tactical survey manuals are up to date and on their apparatus.

D. The fire marshal shall be responsible for reviewing each tactical survey to ensure its accuracy prior to distributing the surveys to all of the companies.

E. The fire chief shall be responsible for identifying the target hazards to be surveyed and for ensuring strict compliance with this standard.

Emergency Operations, SOP 601.01

HIGH-RISE BUILDINGS

I. Scope

This standard lists and defines terms essential to the management of fire control and rescue incidents in high-rise buildings.

II. General

The definition of a high-rise building varies by community. Definitions include more than four stories, more than 75 feet above grade, more than 75 feet above the department's tallest ladder, etc.

III. Definitions

A. Combination wet standpipe system: A system of piping and fire pumps built into a structure to supply water under pressure for firefighting purposes. Fire department connections are provided so that fire department pumpers can supplement either the water supply or pressure. Combination wet standpipes are generally equipped with dry hose at valved outlets. Fire pumps operate on demand.

B. Critical-path construction: A system design used in construction projects to schedule various stages of construction operations for optimum speed of production. Such scheduling reduces interference between crafts and minimizes delays usually inherent at various stages of construction.

C. Dry-standpipe system: A system of piping that includes a fire department connection built into a structure for firefighting. Water for firefighting is supplied by fire department pumpers through an external siamese and piping to fire department hoselines that may be connected to outlets on various floors.

D. Exterior or external exposure: The hazard of ignition to a building or its contents from a fire in an adjoining building or some other exterior source.

E. HAD: An abbreviation for heat-activated device (thermostat).

F. HVAC: An abbreviation for heating, ventilation, and air-conditioning system.

G. Interior or internal exposure: The hazard of ignition to a room or its contents from a fire within the same building.

H. Limited-load area: A surface area that may appear to be solid but that is actually designed to carry certain maximum-weight loads.

I. Nonrequired exit: A means of egress in addition to required exits.

J. Occupancy: The use or function of a building or a portion thereof.

K. Occupant load: The number of people normally occupying a building or floor.

L. Pipe and duct shaft: A vertical or horizontal enclosed passageway housing service utilities, piping, and ducts.

M. Plenum: A container that encloses a volume of gas under greater pressure than the atmosphere surrounding the container—e.g., ductwork.

N. Poke-through construction: A method used to bring service utilities into a building area of a given floor by drilling holes through the concrete floor. Ducts, service pipes, wiring, and the like are connected through these holes to the master utilities in the attic space of the floor below.

O. Programmed elevator: An elevator controlled by electronic devices. These devices automatically schedule stops at various floors to serve the demands of building occupants during periods of peak traffic.

P. Required exit: A legal means of egress for occupants of a building.

Q. Scissor stairways: Two stairways in the same shaft that serve alternate exits or alternate floors. Scissor stairways may or may not include common landings.

R. Smoke tower: An enclosed stairway accessible from each floor only through balconies open to the outside air. Smoke and fire won't normally spread into the smoke tower even though the doors are left open.

S. Stack effect: The accelerated movement by convection of enclosed, heated air, as in a smoke stack or chimney.

T. Structural-load design: The total weight that a structure or portion thereof is engineered to support.

U. Wet-standpipe system: A system of piping that contains water for firefighting purposes and is built into a structure. A wet standpipe is generally equipped with dry fire hose at valved outlets.

Emergency Operations, SOP 601.02

HIGH-RISE OPERATIONS

I. Scope

This standard establishes guidelines and tactical checklists for managing fire control and rescues in high-rise buildings.

II. Incident Command Checklist

Are the following items in place?

Command post

Lobby control.

Interior staging.

Exterior staging.

Rehab sector.

Public information officer.

Crowd/traffic control.

Prefire plans.

Shelters.

Safety zone.

III. Resource Lists

Adequate SCBA/air supply.

Interior communications.

Exhaust fans/lights.

Personnel.

Apparatus.

IV. Do You Have Control of the Following?

Elevator.

Stairwells.

HVAC systems.

Utilities.

Roof access.

Fire protection systems.
Fire pumps.
Alarm systems.

V. Suggested High-Rise Incident Command Organizational Chart:

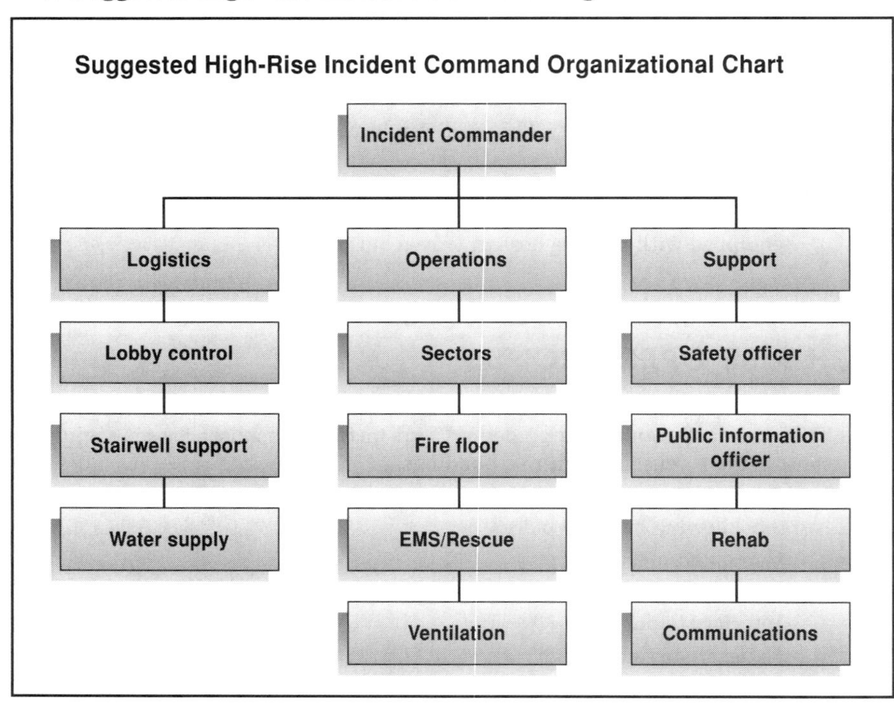

Emergency Operations, SOP 602.01

WATER RESCUE

I. **Scope**

This standard regulates the management of rescue incidents involving persons trapped in bodies of water such as creeks, ponds, and lakes or as a result of flooding or recreational accidents.

II. **General**

A. The safety of the victims and the rescue workers is of paramount importance. Always think before you act. Size-up must be a continuous process.

B. Rescue workers must ensure that they do not become victims themselves.

C. Most water rescues exceed the capabilities of a single engine company. Therefore, never hesitate to summon additional resources.

D. Rescue priorities:

1. Safety of the rescue workers.
2. Safety of the victims.
3. Body recovery.

III. Definitions

A. Eddy current: An area of generally calmer water behind an obstruction in the current. Also, along the shore at corners or rapidly widening areas. A good place to rest or to try to get out of moving water.

B. Eddy fence: A visible line where the current changes to the opposite direction of the main flow and creates an area of eddie current.

C. Hydraulic: Also known as a hole, a keeper, or the drowning machine. This is probably the most dangerous area of moving water and occurs anytime water suddenly drops from a higher to lower level. The most dangerous areas are from two to six feet deep. This can occur over automobiles, rocks, or just about any obstacle in moving water. It is nearly impossible to swim in a hydraulic due to the aeration of the water. Learn to identify hydraulics and avoid them.

D. River left/river right: Used to indicate the appropriate side of a stream. Face downstream to orient yourself to river left or right. This way, stream banks will have the same name no matter who is describing them.

E. Strainer: A buildup of debris that restricts the flow of water. Strainers are very dangerous—avoid them.

F. Upstream/downstream Vs: Upstream Vs point to obstacles. Downstream Vs point to areas of greater water flow.

IV. Risk Techniques

A. The following rescue scenarios are listed in order of their potential risk:

1. Shore-based: Least risky. Use poles, floats, throwbags, etc., to make contact with the victim. Avoid entering the water whenever possible.
2. Tethered boat or float and rope system: A boat or float that is secured at either two or four control points may be used to reach victims. Rope systems range from throwbags to very extensive and complicated tyrolean and two-drag systems.
3. Free boat: A boat that is either paddled or motor powered and not tethered to the shore.
4. In-water contact: Swim to the victim, physically contact him, and return him to the shore. *This is very risky!*
5. Helicopter: The final resort when other means have failed or are obviously impossible.

B. Basic safety rules:

1. Wear a personal flotation device in or near the water. This is mandatory.
2. Wear a rescue helmet in the water.
3. Do not wear structural protective clothing in or near the water.
4. Do not overextend yourself.
5. Do not become the victim.
6. Always bathe thoroughly after entering any body of water due to the pollution hazard.

C. Equipment:

1. Personal flotation device (PFD): Must be worn at all times during a water rescue when in or near the water. Inspect PFDs for rips, tears, flotation compartments that leak, etc., prior to each use.
2. Helmet: It is strongly advised that the rescuer wear an SWR helmet

anytime he is in the water. Inspect the helmet for cracks, loose padding or straps, and any other abnormal condition prior to each use.

3. Rope: Should be clean and dry before storing. Rope that is used for water rescue should never be used for any other purpose.
4. Knives: Should be securely fastened to the shoulder strap of the PFD of the rescuer entering the water. These knives should only be used in water rescue operations.

Emergency Operations, SOP 603.01

EMERGENCY VEHICLE PLACEMENT

I. **Scope**

This standard regulates the placement and positioning of emergency vehicles at fire control and rescue incidents.

II. **General Rules for Positioning Emergency Vehicles**

A. Apparatus function shall regulate placement. The first-arriving companies should position themselves to maximum advantage and go to work. Later-arriving units should place themselves in a manner that builds on the initial plan and allows for expansion of the operation. However, at no time shall apparatus be positioned in a manner as to place it in an unnecessarily dangerous position.

B. Apparatus that is not being used should be staged in an uncommitted position that will not congest the incident site and will facilitate performing a maximum number of evolutions. Likewise, private vehicles brought to the scene by off-duty or support personnel should:

1. Not block the street. Leave the street open at all times to permit the movement of additional apparatus.
2. Park at least one block away and never in a cul-de-sac or in front of the building(s) involved in the incident.
3. Do not park in front of fire hydrants, automatic sprinklers, or standpipe connections.
4. Do not block private driveways.
5. Do not run over or park on fire hose with your vehicle.

C. When placing or positioning an emergency vehicle at an incident, consider the following items:

1. Leave an access lane open down the center of the street.
2. Do not park in such a manner as to make the emergency vehicle an exposure.
3. At fires, avoid heat and smoke. Guard against possible building collapse.
4. Do not become trapped. Allow adequate room to reposition the vehicle if necessary.
5. Beware of overhead power lines.
6. Try to use hoselines and equipment off of apparatus at the immediate scene to maintain better control and not strip all the apparatus.
7. On multiple alarms, an officer shall be assigned to stage apparatus and deploy them per the incident commander's request.
8. Take maximum advantage of key hydrants to avoid excessive lays and to ensure maximum use of the water supply.

9. Do *not* park too close to another emergency vehicle. Allow sufficient room to deploy hoselines and to remove stretchers, ladders, and other equipment from the vehicle.

D. The first-arriving engine company at a fire should normally proceed just past the front of the fire, laying a supply line if necessary. ("Front" does not necessarily mean the front of the building. It may mean the area of heaviest involvement or primary access.) Driving past the structure this way will enable its crew to see three sides of the building.

E. Position the engine to use its deck guns, floodlights, etc. Take care to leave room for ladder or support companies. If the building has a wide frontage, position the engine at the entrance that provides the best access to the fire.

F. Unless otherwise directed, the second-arriving engine company at a working fire should proceed to the rear or secondary access point and go to work.

G. If nothing is showing, the second-arriving engine should stage at the nearest hydrant and await orders if the first-arriving company did not lay hose.

H. Truck companies should initially stage in such a position as not to congest the incident scene. A truck company should position itself for maximum use as ordered by the incident commander.

I. Additional arriving companies should stage at least one block away and request orders from the incident commander or the staging officer. Companies should commit only when ordered to do so. These companies should report the number of personnel in their respective companies, as well as their staging location. Example: "Engine Company Two staged one block south with three personnel." Staged companies should check their map books to locate key hydrants, etc.

III. Medical Incidents

A. If an engine or truck company arrives before the medic unit, the company should leave clear access for the medic unit and not block the incident site.

B. Due to the danger of oncoming traffic at motor vehicle accidents, the engine or truck should be parked so as to provide a barrier for personnel.

IV. Staging

A. Level I Staging.

1. First engine to scene.
2. Second engine and additional engines—one block away and report location. Check map books.
3. First truck company to scene.
4. Privately owned vehicles—park at least one block away, not blocking the street, driveways, or hydrants.
5. Level I staging begins on arrival of the first company. The first company gives a situation report and advises the action to be taken.
6. Additional companies advise staging location.
7. If additional companies lay, report so to Dispatch.

B. Level II Staging: Multiple Alarms/Mass-Casualty Incidents

1. Command designates the staging area and staging officer.
2. All companies and personnel report to the staging area and await assignment.
3. POVs report to one block of staging area.
4. Dispatch shall designate a channel for staging and inform Command.

5. "Staging" shall be used as the radio identifier. Staging maintains a log of the personnel and apparatus available.
6. Staging should maintain a reserve of apparatus and personnel unless instructed otherwise.

C. Level III Staging: Multiple Alarms/Mass-Casualty Incidents
1. To be used when severe weather makes Level II staging inadvisable, such as during floods, tornadoes, and winter storms.
2. Companies will back in at a fire station or other covered facility and await orders.
3. Privately owned vehicles shall not obstruct the staging area.
4. Dispatch shall designate a channel for staging and shall report the status to Support or Command.
5. "Staging" shall be used for a radio identifier. Staging shall maintain a log of the personnel and apparatus available.
6. Staging should maintain a reserve engine and ladder company in the department affected. Use move-ups.

Emergency Operations, SOP 603.02

PERSONNEL DEPLOYMENT ON THE FIREGROUND

I. <u>Scope</u>

This standard regulates the deployment of personnel at the scene of a fire control incident.

II. <u>General</u>

A. It shall be the policy of the department to dispatch a minimum of _____ (recommended minimum of 16) personnel to a reported structure fire on the initial alarm. Incident volume may occasionally reduce the number of personnel available to respond. In such cases, the incident commander should alter the tactical operations to ensure adequate firefighter safety.

B. To assemble the recommended minimum number of personnel at an incident, it will normally be necessary to dispatch ____ engine(s), ____ truck or rescue squad(s), and ____ ambulance(s) on the initial alarm.

III. <u>Offensive/Interior Operations</u>

A. Assembling 16 personnel at a fire control incident will allow for an aggressive interior attack with a maximum fire flow of 500 gpm, and it will also enable the proper support functions to be performed.

B. Personnel should be deployed as follows:

Command	1
Pump operator	1
Search and rescue	2
Ventilation, forcible entry, and/or salvage	2
Hoselines (two)	6
Paramedics—aid station, rehab, and safety	2
Rapid intervention team (RIT)	2
TOTAL FIRST-ALARM ASSIGNMENT:	16

C. The accompanying illustration shows the possible deployment of the initial-alarm assignment for a typical room-and-contents fire in a single-family dwelling. Obviously, it may be necessary to alter the deployment scenario depending on the actual situation.

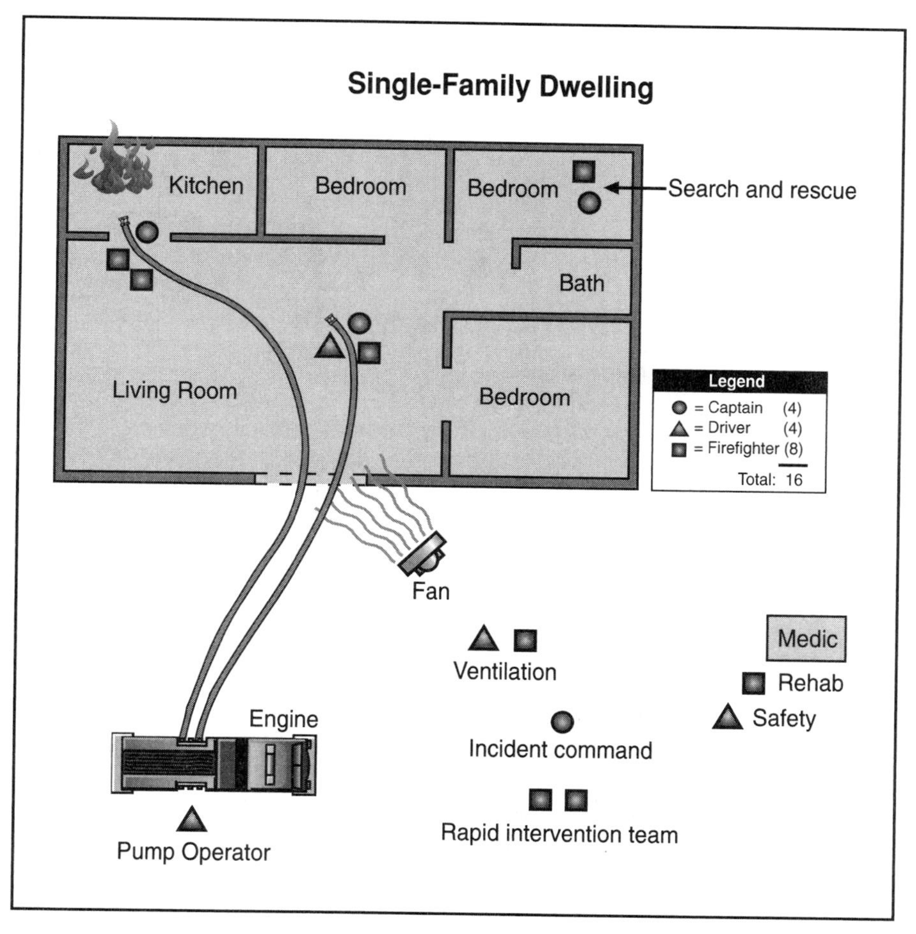

D. Individual Company Assignments

1. First-arriving engine company:
    a. Proceed past the structure to size up its three visible sides and then position appropriately.
    b. Lay supply line if necessary or notify another engine to bring one in.
    c. Place the first attack line into service to support containment, rescue, and exposure protection. Always attack from the unburned side.
    d. At night, place exterior lighting in service and illuminate the front and sides of the building.

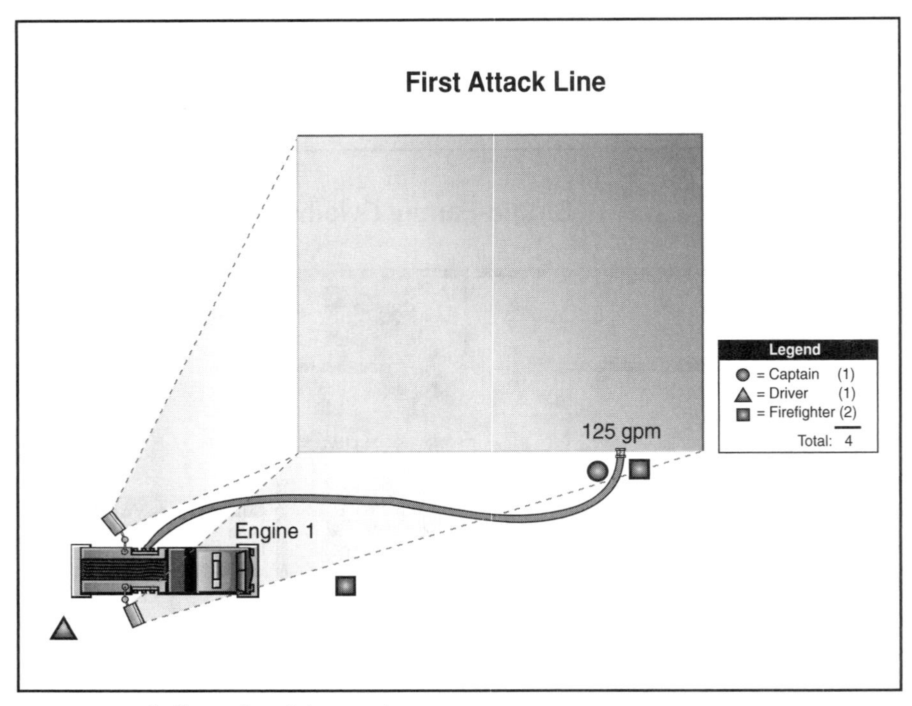

2. Second-arriving engine company:
   a. Size-up and placement:
      (1) Level I staging at nearest hydrant.
      (2) Lay supply line if required.
      (3) Go to rear of building if possible.
      (4) Place a second line in service as a backup to the first line, but *do not* oppose the first line.
      (5) Assist rescue.
      (6) Cover secondary exposure.
      (7) Supplement exterior lighting.
      (8) Establish RIT.

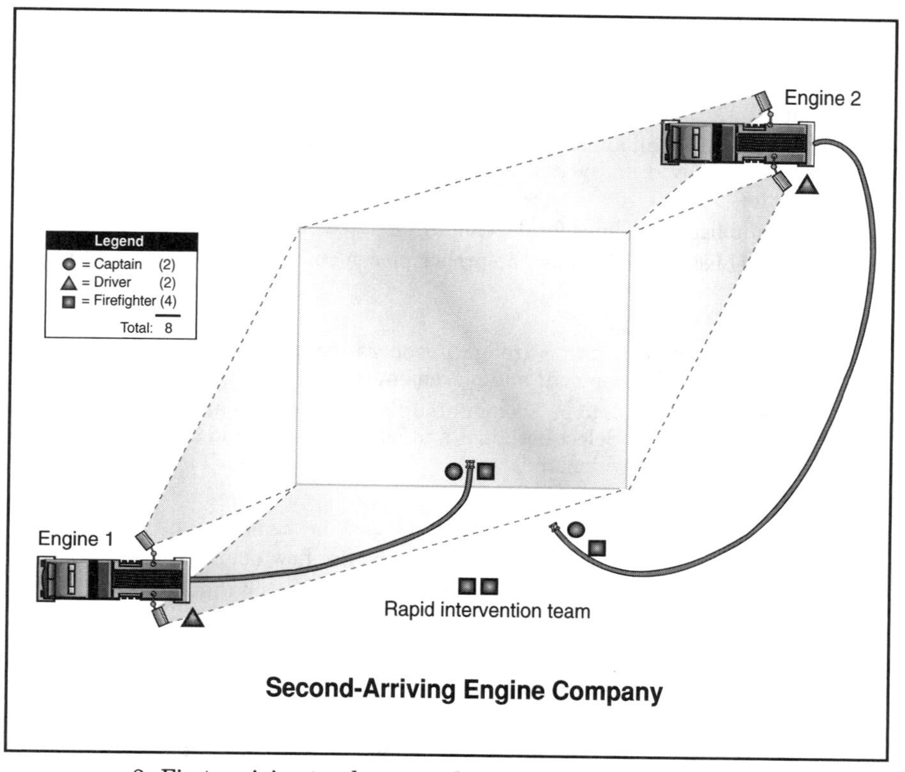

Second-Arriving Engine Company

3. First-arriving truck or squad company:
   a. Size-up and placement:
      (1) Front or front corner of the building.
      (2) Ladder as required.
   b. Search and rescue:
   c. Ventilation:
      (1) Positive pressure.
      (2) Natural: Open roof as ordered by Command.
4. Third-arriving engine:
   If *not* ordered to lay:
   a. Level I stage, do not block street.
   b. Report number of personnel to the incident commander.
   c. Report to scene with SCBA, portable radio, pike pole, ax, and salvage covers.

Emergency Operations, SOP 603.03

FIRE STREAMS

I. **Scope**

This standard shall apply to all fire control and rescue incidents where hoselines must be deployed and water is used to mitigate the situation. It was promulgated to:

A. Establish guidelines for developing effective fire streams.

B. Establish guidelines for the proper placement of fire streams at a fire control or rescue incident.

II. **General**

A. Tactical success during firefighting operations depends to a large degree on the timely development and placement of adequate fire streams.

B. For a fire stream to be adequate, sufficient pressures are needed to provide effective reach. Select hoselines and nozzles that provide sufficient volumes to meet fire flow demands.

III. **Hoseline Selection**

A. Initial hoselines should be deployed based on the following:

1. Hose of a sufficient size to meet the fire flow demand.
2. The proper safety factor to personnel operating the line.
3. Adequate personnel available to deploy the line effectively.
4. Sufficient water supply to meet fire flow demand.
5. The time required and work involved in deploying stream.

B. Stream comparisons:

1. Initial hose streams will commonly use preconnected lines of either (insert preferred sizes) _______ or _______.
2. 1¾-inch lines will normally be equipped with (insert nozzle type) ___________. At 100 psi, flows of 95 to 200 gpm are available.
3. 2½-inch lines will normally have (insert type) ___________ fog nozzles. At 100 psi, flows of 120 to 250 gpm are available.
4. Standpipe lines commonly have 95-gpm nozzles.
5. For defensive operations, smooth-bore 2½-inch nozzles are carried on all engine companies. Tips of 1 inch provide 210 gpm and 1⅛-inch tips provide 266 gpm at 50 psi.
6. Deck pipes have stacked tips and at 80 psi are capable of delivering:

   1⅜-inch—502 gpm
   1½-inch—598 gpm
   1¾-inch—814 gpm
   2½-inch—1669 gpm

7. Portable master stream devices are equipped with both straight tips and 500-gpm fog nozzles.
8. (Insert type) _________ fog nozzles provide 350 to 1,000 gpm at 100 psi.

IV. **Placement**

A. Order of deployment:

1. Place the first line between the fire and the victims to provide an escape route.

2. If there is no life hazard, place the first line between the fire and the most severe exposure.
3. Place the second line to support the first line or to protect a secondary means of egress.
4. Use additional lines to support attack positions.

B. Rules of placement:
1. Always attack structure fires from the unburned side.
2. The safest attack route in wildland fires is from the burned side.
3. Never combine interior and exterior attacks.
4. Use the direct attack method in interior operations—i.e., apply water on the base of the fire using 30° fog streams. (Note: Some departments prefer smooth-bore nozzles for interior attack.)

V. **<u>Defensive Fires/Exterior Attack</u>**

A. Large fires require large volumes of water. Whenever possible, avoid using handlines and opt for portable monitors and deck pipes instead.

B. The best results are normally obtained from smooth-bore nozzles. They provide greater reach and penetration. Fog lines may be necessary in some cases to protect apparatus and personnel.

C. Elevated streams:
1. Use to protect exposures.
2. Use through windows.
3. Never use in vent holes.

Directing an elevated stream through a properly made ventilation hole in a roof will negate the beneficial results of the vent. The stream will act as a cover that arrests ventilation, drives fire to other areas, and often forces firefighters out of the building.

D. The following table provides friction losses in elevating platforms.

Pressure Loss (psi) Due to Elevation

Height of nozzle (feet)	55	65	75	85	90	125
Head loss—Rule of thumb (5 pounds per 10 feet)	27.5	32.5	37.5	42.5	45.0	62.5
Actual (.434 pounds per ft.)	23.9	28.2	32.6	36.9	39.1	54.3
Pressure loss in 4" pipe (750 gpm)	22.0	22.0	22.0	22.0	22.0	22.0
Pressure loss in 4" pipe (1,000 gpm)	37.5	37.5	37.5	37.5	37.5	37.5
Total loss—4" (thumb) (750 gpm)	49.5	54.5	59.5	64.5	67.0	84.5
(actual)	45.9	50.2	54.6	58.9	61.1	76.3
Total loss—4" thumb) (1,000 gpm)	65.0	70.0	75.0	80.0	82.5	100.0
(actual)	61.4	65.7	70.1	74.4	76.6	91.8

Emergency Operations, SOP 603.04

HOSE LOADS

I. Scope

This standard regulates the loading and transporting of fire hose on apparatus.

II. General

A. For the purposes of this standard, a hose load may refer to both the total complement of hose carried on an individual apparatus and to the manner in which the hose is carried on an apparatus.

B. There will be some variation in both the amount of hose carried on an apparatus and the manner in which it is loaded due to number, size, and arrangements of the various hosebeds on apparatus built by different manufacturers.

III. Definitions

A. Attack line: Any hose that is manually deployed and supplies water from an apparatus or standpipe discharge to a nozzle used to control or extinguish fire.

B. Combination lay: A hoselay in which hoselines are laid in both directions: forward (from a water source to a fire) and reverse (from a fire to a water source).

C. Filler hose: A section of 2½-inch or 3-inch fire hose carried on an engine and used to fill the booster tank from a hydrant or other water source.

D. Forward lay: Advancing or deploying a supply line from a water source to the fire or incident scene.

E. Hard suction: A 10-foot length of noncollapsible, rubberized hose with a steel core that is used for drafting.

F. Hosebed: An area or compartment on an apparatus designed to carry or transport hose.

G. Hose book: A journal that is carried on an apparatus and that lists the current hose inventory by section number, location on the apparatus, and date loaded. It is the company officer's responsibility to maintain the hose book and make sure all information is current and accurate.

H. Hose carry: A method of moving and deploying fire hose.

I. Hose pack: A compact bundle of hose normally used in standpipe operations.

J. Hose rack: A portable or fixed storage unit for fire hose.

K. Hose record: A permanent record provided for each section of hose listing the history of an individual section of hose from the time of purchase until it is taken out of service. It is the responsibility of the hose officer to maintain the hose record.

L. Preconnect: An attack hose connected to a discharge when the hose is loaded so as to reduce deployment time.

M. Reverse lay: A method of laying hose from a fire or incident scene to a water source.

N. Service test: Hydrostatic pressure testing of fire hose conducted at least annually.

O. Supply line: Hose from a water source or supply to the intake of an apparatus, sprinkler system, or standpipe.

IV. Supply Lines

A. The department uses large-diameter hose (LDH) supply lines on all front-line engine companies. Each engine should carry a minimum of _____ feet of _____-inch hose coupled in _____-foot sections. The hose is carried on the apparatus in a flat load. Single sections of 100′, 50′, 25′, and 10′ are also carried in rolls to facilitate deployment.

B. Reserve engines may carry _____ feet of _____-inch supply line in lieu of LDH. The _____-inch hose is coupled in _____-foot lengths. This hose is loaded flat and is divided into two beds of _____ feet each. The hose shall be loaded for a forward lay (i.e., the last coupling on top will be female) and the two beds shall be connected to facilitate laying a single line.

C. Each engine shall carry a minimum of _____ feet of _____-inch supply line. This hose is used to supply portable master stream appliances, standpipes, and automatic fire sprinkler systems.

V. Attack Lines

A. The number and lengths of attack lines vary by apparatus due to the limitations imposed by the number and sizes of hosebeds found on a given apparatus.

B. Preconnects: _____-inch.

1. Loaded in __________ load.
2. Color coded: (Note: The outer jacket of the hose is the color indicated below and corresponds to the color code of the individual discharge outlets.)
   a. Standpipe pack—orange.
   b. 100 feet—yellow.
   c. 150 feet—blue.
   d. 200 feet—green.

C. Preconnects: (Insert size—e.g., 2½-inch)

1. Each engine company will normally carry a _____-foot preconnect of _____-inch hose.
2. The hose is coupled in _____-foot sections and is loaded in a ________ load.

D. Preconnects: One-inch line

1. Wildland apparatus will carry a minimum of _____ feet of 1-inch redline on a reel.
2. A short section of 1-inch redline may also be carried on the front bumper.

VI. Forestry Hose

A. Each engine company and wildland apparatus shall carry a minimum of _____ feet of 1-inch forestry hose. Forestry hose is coupled in 50-foot sections.

B. (Reserved)

VII. Hard Suction

A. Engine companies shall carry a minimum of _____ sleeves of hard suction hose for drafting purposes.

B. When drafting, a strainer shall always be used.

Emergency Operations, SOP 603.05

WATER SUPPLIES

I. Scope

This standard establishes the requirement that an adequate and reliable water supply be established at each incident.

II. General

A. For firefighting efforts to be effective, an adequate and reliable supply of water must be available. The adequacy and reliability of potential sources of water are constantly changing due to weather, system demands, and many other factors beyond the department's control.

B. Each member shall be aware of the potential fire flow demands within his district and shall identify available options for developing a sufficient volume of water to adequately combat any fire that might occur.

III. Responsibilities

A. The fire chief shall appoint a member to serve as the water supply officer for the department.

B. The water supply officer shall:

1. Serve as a liaison with the water department.
2. Furnish an accurate and up-to-date hydrant map to each fire station and fire company.
3. Maintain a complete and up-to-date water supply map at Administration/Headquarters.
4. Identify areas where additional hydrants are needed and work with the water department to have them installed.
5. Provide each station with blue hydrant markers and supervise their installation to facilitate locating hydrants during emergencies. This effort shall be coordinated with the street department to ensure that markers are replaced after repaving projects are completed.
6. Provide each station and company with a complete and accurate map identifying static water sources. The map shall provide information concerning accessibility and capacity for firefighting.
7. Supervise fire flow testing and record the results. This should be performed in cooperation with the fire protection engineer.
8. Maintain and distribute an up-to-date list of hydrants that are out of service.

C. The tactical survey officer shall have overall responsibility for developing tactical surveys for all schools, institutional occupancies, high-rises, and target hazards. A master copy shall be placed in the occupancy's inspection file and a copy of *all* tactical surveys shall be carried on all command vehicles, engine companies, and truck companies. Fire flow demand and water supply information shall be included on the tactical surveys.

D. The shift training officers shall conduct monthly territory drills with the members of their shifts to acquaint members with the location and capacity of the fire hydrants and static water supplies within the city, county, or district. Water supply information shall be included on the monthly territory tests administered by the STOs.

E. Company officers shall be responsible for:

1. Knowing the correct location and capacity of each fire hydrant and static water source within their first-due area.
2. (Recommendation; actual practice will vary.) Installing hydrant markers on the streets within their assigned districts to assist in locating fire hydrants at night.
3. Conducting an annual test of each hydrant within their districts to ensure that they are working properly.
4. Reporting high grass, weeds, and other hydrant obstructions to the fire marshal's office for correction.
5. Maintaining an up-to-date list of hydrants that are out of service within the district.

F. Drivers shall be responsible for knowing the correct location and capacity of each fire hydrant and static water source within the district. Drivers shall also be responsible for ensuring that a street and hydrant map is maintained in their assigned apparatus and that the map is current and up to date.

IV. Hydrant Color Codes

A. Hydrants are color-coded in accordance with NFPA 291, *Recommended Practice for Fire Flow Testing and Marking of Hydrants,* as follows:

Color	Flow
Red	≤ 500 gpm
Orange	500-1,000 gpm
Green	1,000-1,500 gpm
Blue	≥ 1,500 gpm

B. Whenever a hydrant is discovered that has not been color-coded, it shall be reported to the water supply officer (WSO). The WSO shall request that the water department properly code the hydrant.

V. Operational Procedures

A. Each engine company shall be responsible for providing its own uninterrupted water supply on the fireground. The ability to do so will be predicated on:

1. The required fire flow.
2. The available water supply.
3. The number of personnel available.
4. The numbers and types of available apparatus.

B. Calculations of required fire flow:

1. The following factors influence the required fire flow:
   a. Construction type.
   b. Contents.
   c. Occupancy.
   d. Exposures.
   e. The presence or absence of extinguishing systems.
2. For tactical surveys, the calculated analysis of the fire flow shall use the following Insurance Services Office (ISO) formula:

   $Q = 18\ C\ (A)^{.05}$

   Q = Needed fire flow in gallons per minute (gpm).

   A = Total building area in square feet (ft^2).

C = A factor based on construction as follows:

- C = 1.5 for wood frame.
- C = 1.0 for wood-joisted masonry.
- C = 0.8 for unprotected noncombustible.
- C = 0.6 for fire resistive.

3. For a quick reference during a fireground operation, use the Nelson-Royer Formula for the required fire flow:

$$\text{Required fire flow (gpm)} = \frac{H \times W \times L}{100}$$

- H = Building height.
- W = Building width.
- L = Building length.

Example: For a building that is 60 feet long and 40 feet wide with a 10-foot ceiling, the fire flow would be:

$$\text{gpm} = \frac{10 \times 40 \times 60}{100} = \frac{24{,}000}{100} = 240$$

To flow 240 gpm would require:

- One 2½-inch handline flowing 240 gpm or
- Two 1¾-inch handlines at 120 gpm each.

Note: The Nelson-Royer Formula should only be used for the calculation when the ISO formula hasn't been computed on the tactical survey.

4. As a general rule of thumb, the following minimum flows are required:

Residential	500 gpm
Light commercial	1,000 gpm
Heavy commercial	1,500 gpm
Industrial	≥ 2,000 gpm

5. The required fire flow may be reduced by 50 percent if an automatic fire sprinkler system is present.

C. Water supplies may be established by:

1. Booster tank operations: For demands of less than 250 gpm or brief duration. Water may be transferred from other apparatus by 2½-inch lines. If more than 1,000 gallons are needed, use another source.
2. Supply lines: A 5-inch line from a hydrant or static source via drafting. This is the most dependable source and should be used whenever possible. When in doubt, lay a line.
3. Tender shuttle: The use of water tenders (aka tankers) to transport water in areas without hydrants. Tenders are available from (insert sources):

   ____

   ____

4. Dump and pump: (1) First, two members set up a dump-and-pump tank. (2) Then, a tanker or auxiliary water truck fills the dump-and-pump tank. (3) A pumper can draw water from the tank even while it is being filled. (4) Once the tanker is empty, it can go for more water while the pumper continues to draw from the tank.

Emergency Operations, SOP 603.06

VENTILATION

I. Scope

This standard applies to incidents involving structures and confined spaces where the prompt removal of smoke, heat, and other products of combustion is necessary to quickly and safely extinguish the fire.

II. General

A. Prompt and efficient ventilation is necessary to mitigate the potentially harmful effects of smoke, heat, and other contaminants within structures and confined spaces.

B. If unchecked, smoke and heat contribute to property damage and can injure and kill those who become trapped.

C. Smoke and heat also hinder firefighters in their efforts to perform search and rescue operations as well as suppression.

D. Therefore, it is the policy of this department to provide prompt and proper ventilation in all buildings and confined spaces in which smoke, heat, or other products of combustion are present unless otherwise ordered by the incident commander.

III. When to Ventilate

A. Ventilation shall always be performed whenever:

1. Heat, smoke, and other products of combustion are present.
2. Hose crews cannot effectively make an interior attack due to excessive heat and poor visibility.
3. Heat, smoke, and other products of combustion block escape routes for the occupants of the structure.

B. When performing ventilation, the following safety precautions should be observed:

1. Read the smoke. Observe conditions that might indicate that the potential for flashover or backdraft is present.
2. *Never* direct hose streams into ventilation openings.
3. Always have charged hoselines in place prior to beginning ventilation.
4. Maintain communications.
5. Wear full protective clothing and SCBA.
6. Always consider structural soundness.
7. Exercise caution whenever using power saws, axes, and other sharp instruments.
8. Secure a lifeline to any firefighter who is on a potentially weakened roof.
9. Remember that improper ventilation techniques may contribute to fire spread.

IV. Types of Ventilation

A. Natural: Accomplished by making use of wind currents. Open the building on the leeward side to allow the smoke to escape, then open the windward side to provide fresh air currents.

B. Mechanical: Use of electric or gasoline-powered fans or blowers to evacuate smoke from a building or confined space.

1. Negative pressure: Exhaust smoke from the building.
2. Positive pressure: Blow fresh air into the building to force the smoke out.

C. Horizontal: Generally inflicts less damage to the building than vertical ventilation, since it is typically accomplished through available portals such as doors and windows.

D. Vertical: May also take advantage of natural building features, such as skylights, shafts, and rooftop stairways. Many times, the only option is to cut ventilation openings into the building itself.

Emergency Operations, SOP 604.01

SUPPORT OF AUTOMATIC SPRINKLER SYSTEMS

I. Scope

This standard regulates emergency operations in buildings equipped with automatic fire sprinkler systems. It was promulgated to:

A. Establish guidelines for emergency operations in buildings equipped with automatic fire sprinkler systems.

B. Ensure that automatic fire sprinkler systems are properly supported so that they perform as designed.

C. Ensure that activated automatic fire sprinkler systems are restored to service properly and that the fire has been extinguished.

II. General

A. It shall be the policy of this department to support and supplement automatic fire sprinkler systems that have activated during a fire.

B. It shall be the responsibility of all personnel to know which buildings within the city, county, or district are equipped with automatic fire sprinkler systems and to be familiar with the location of fire department connections and control valves.

C. Each fire company shall be provided with an updated list of automatic fire sprinkler systems and shall carry the list on their apparatus.

D. Companies shall identify whether systems are pipe schedule, hydraulically calculated, 13D, or 13R. This shall be noted on the sprinkler list. (See NFPA 13, 13D, and 13R for definitions and details. The type of system will affect flow and pressure requirements.)

III. Operations

A. System support:

1. The second-due engine company should lay a supply line to the fire department connection (FDC) and then supplement the system by pumping two 2½-inch or 3-inch lines. It is best to connect to both inlets of a sprinkler siamese, since doing so decreases friction loss. It also provides safety and redundancy if one of the lines bursts.

   Note: Some sprinkler systems have only a single 2½-inch or 1½-inch inlet.

2. Pumping pressures:
   a. Pipe schedule: 150 psi at the fire department connection plus 5 psi per floor in multiple-story buildings.
   b. Hydraulically calculated systems and 13R or 13D systems: 100 psi

at the fire department connection. The goal is to provide 50 psi at the most remote head.

c. More water may be required depending on the number of heads operating, building size and height, etc. Engine companies shall adjust pressures accordingly.

3. The engine company supporting the system shall not be used for handlines or other similar operations.

IV. System Restoration

A. The system should not be shut off until the fire is extinguished, and only then by order of the incident commander. The member shutting off the system shall remain at the control valve until relieved by the incident commander.

B. Prior to shutting off the system, the water flow may be stopped by using sprinkler wedges. Each firefighter and company shall be issued sprinkler wedges.

C. Before leaving the scene, ensure that the system is placed back into service.

Emergency Operations, SOP 604.02

STANDPIPE OPERATIONS

I. Scope

This standard applies to the management of incidents in buildings equipped with standpipe systems.

II. General

A. It shall be the policy of this department to use standpipe systems, when provided, to support interior firefighting operations. In buildings equipped with standpipes, the following operations shall be performed:

1. At least one engine company from the initial-alarm assignment shall lay a supply line and support the standpipe system by pumping at least two 2½-inch lines into the standpipe connection. A minimum of 150 psi should be supplied to the standpipe connection for elevations under 100 feet. Add five psi for each additional floor above 100 feet.
2. All interior operations shall be conducted using fire department hose. The first line shall be connected below the fire floor. Additional lines shall be added as needed.
3. A check shall be made to ensure that all water supply valves are open and that the fire pump, where provided, is operating properly. A secondary check shall be made to ensure that all hose outlets not in use are closed.
4. The incident command system shall be used. An interior sector shall be established and sectors shall be assigned by floor numbers or other identifying areas such as the roof or lobby. Communications shall be established among all sectors.
5. A lobby control shall be established at the point of entry and no unauthorized persons shall be admitted. The names of all persons entering shall be recorded as well as their time of entry and exit.
6. In high-rise buildings, an interior staging area shall be established on a floor below the fire floor. Reserve personnel and equipment shall be assembled and shuttled to crews operating on the fire floor.
7. At minimum, the initial attack crews shall take the following equipment into the building:

a. Appropriate protective equipment and SCBAs.
b. Standpipe hose packs along with the appropriate adapters and spanner wrenches.
c. At least one ax, one pike pole, one pry bar, rope, handlights, portable radios, and any other equipment the company or incident commander may deem appropriate.
d. Equipment assignment:
1) Officer: SCBA, portable radio, handlight, ax.
2) Driver: SCBA, rope, pike pole.
3) Firefighter: SCBA, hose pack, pry bar.

Other equipment might be taken as the crew size varies or as ordered by the incident commander.

8. The incident commander shall immediately cause proper ventilation and lighting operations to commence to adequately support interior rescue and firefighting efforts.
9. As soon as possible, efforts should be made to provide an adequate and continuous supply of air for all SCBAs.
10. All other operations shall be conducted per established SOPs and tactical guidelines.

Emergency Operations, SOP 605.01

AIRPORT RESPONSE

I. Scope

This standard shall regulate responses to aircraft incidents at municipal, county, or district airports. This SOP is written for airports that do not have on-site crash stations.

II. Response Procedures

A. Dispatch operators shall obtain the following information, whenever possible, prior to dispatching apparatus to an incident involving an aircraft at an airport:
1. The nature of the emergency.
2. The exact location if the aircraft is down. If the aircraft is aloft or did not make it to the airport:
a. Its distance from the airport and its ETA.
b. Its direction of travel.
3. The size of the aircraft.
4. The number of passengers and the nature of the cargo.
5. The equipment required.
6. If a military aircraft, the type of ordnance on board.
7. Other pertinent data.

B. Response:

The following equipment shall be dispatched to an aircraft emergency:
1. The closest engine company with foam and extrication equipment. In the event of an actual crash, dispatch a full first-alarm assignment.
2. The nearest available ambulance.

3. Respond a back-in (also known as a fill-in or standby) to the station nearest the airport. This engine will respond to the airport in the event of a crash if requested by the incident commander.
4. The emergency management coordinator shall be dispatched to all actual crashes.
5. Responses for standbys are to be nonemergency if the aircraft is more than 15 minutes away from the airport.

C. Staging:

1. Stage emergency vehicles on the aircraft parking ramp, remaining well clear of the runway area.
2. If the emergency occurs at night, turn off the warning lights while the emergency vehicles are stopped in the staging area. Flashing lights can be confusing and disorienting to the pilot, especially in fog or haze.

D. Command:

1. The incident commander shall respond to the terminal and establish a command post there. (The first-due engine should have a key to the terminal building.) The incident commander shall use the airport's radio to communicate with the aircraft. The following frequencies shall be used:

<u>RADIO FREQUENCY LIST</u>

Airport (tower)	____ mHz
Regional Flight Service Station	122.0 or 122.2 mHz
Emergency	121.5 mHz

2. The regional Flight Service Station (FSS) will coordinate with all other agencies. Call the FSS first on 122.2 or 1-800/992-7433. If the aircraft is on the runway, notify the FSS and close the airport.

E. Aircraft radio procedure:

1. First, do not use complicated terminology. All aircraft communication is done in plain English.
2. Express all numbers by individual digits. For example, to transmit the number 4359, say each number separately: *four, three, five, niner*. The number nine is pronounced *niner* so that the pilot can clearly distinguish it from the number *five*. Even if a number is only two digits long, you must say the digits separately. For example, 17 would be said *one, seven,* not *seventeen*.
3. Express all letters per the ICAO (International Civil Aviation Organization) phonetic alphabet (see SOP 801.01). An aircraft bearing the tail number N483PG would be addressed as *four, eight, three, papa, golf* on first contact, and *three, papa, golf* thereafter. The prefix N of an aircraft's tail number is an indicator of United States registry and may be omitted from all transmissions.
4. 10 codes and CB jargon have absolutely no meaning to a pilot.
5. The fire department's role during an aircraft emergency is to provide the pilot with any information he may request or information you feel he needs to know about the airport or the surrounding area. This will help the pilot make the best possible decision as to whether to land there or to divert.
6. Remember, the pilot is the final authority as to the outcome of the aircraft's flight. *We do not have the authority to tell a pilot what to do with his aircraft*. If the aircraft crashes, then we are in command of the scene.

7. If the call for help comes in the middle of the night, we may have to provide the pilot with information concerning the direction of the wind, its general speed, and the runways available at the airport. Example: "Skyhawk 359, the wind is southwest at 15 knots and the runways available are 17 and 35."
8. The pilot only needs to know the general direction (magnetic, not true) that the wind is blowing from to help him decide which runway to use: North, Northeast, East, Southeast, South, Southwest, West, or Northwest. Since runways are named for their orientation with respect to magnetic north, not true north, surface wind information transmitted to a pilot must also be based on its magnetic direction. If the winds are being estimated based on true north, the pilot must be advised so that he can take into account the local magnetic variation in selecting his target runway.
9. The wind speed can be estimated by observing the wind sock located somewhere on the airport. If the wind is blowing at 15 knots or faster, then the wind sock will stand straight out. A 10-knot wind will hold the wind sock about midway from horizontal.
10. If the pilot requests services that your department cannot provide, such as high-expansion foam on the runway, either provide him with the radio frequencies for those who can provide the necessary emergency equipment or attempt to locate the equipment for him.

III. Your Municipal Airport Data

A. The following information applies to your airport:

Airport elevation:	__________ feet.	
Runways available:	__________	__________
Length:	__________ feet.	__________ feet.
Width:	__________ feet.	__________ feet.

B. (Insert an airport map, if available.)

IV. General Information

A. Almost all small single-engine and twin-engine airplanes have a cabin door on the right side. Small high-wing airplanes usually have doors on both sides. Larger twin-engine aircraft and small business jets have an entrance door on the left side. Small helicopters tend to have doors on both sides. Large helicopters vary according to purpose.

B. All airplanes carry fuel in the wings. Some have fuel tanks in the belly.

C. While working around a propeller-driven aircraft, *do not* stand within the arc of the propeller. If the propeller is rotated only part of a revolution, the engine may fire, causing severe injury.

D. Note: Some jet aircraft use an acid-based hydraulic fluid that is caustic—use caution.

Emergency Operations, SOP 605.02

MOTOR VEHICLE FIRES

I. Scope

This standard was promulgated to ensure the safety of members involved in suppressing motor vehicle fires.

II. General Guidelines

A. The number and variety of motorized vehicles increases each day. It is impossible to completely identify all of the problems that might be encountered while extinguishing such a fire. Nevertheless, some commonalities do exist, and the general guidelines of this standard should be followed when combating a fire in a motorized vehicle.

B. The incident commander should always remain cautious and assume the worst until he can be certain that conditions are safe. For example, if a burning vehicle is placarded as transporting a hazardous material, assume that the incident is a haz-mat incident until it can be ascertained that no haz mats are burning or have been released.

C. If the fire appears to be the result of an accident, the incident commander should consider the mechanism of the accident in developing his strategy for managing the incident.

D. The first priority at the scene of a vehicle fire is rescue. The incident commander must assume that someone is trapped in the burning vehicle until having ascertained that all of the occupants of the vehicle have gotten out.

E. The second priority is the safety of the firefighters, rescue workers, and spectators. If spectators are present, remove them to a safe distance. Do not place firefighters and rescue workers in harm's way unnecessarily. Remember, do not risk a lot for a little.

F. It is important to control the flow of traffic. The need to minimize the disruption of traffic in heavily congested areas must be balanced by the need to provide for the safety of firefighters and rescue workers. Whenever possible, direct traffic away from the incident.

III. Procedures

A. Apparatus placement:

1. Position apparatus upwind and uphill from the burning vehicle. This will keep the crew out of the smoke and prevent leaking fuel from running underneath the apparatus.
2. Apparatus should not be parked closer than 100 feet from the burning vehicle whenever traffic conditions permit. If the burning vehicle is labeled as transporting a hazardous material, increase this distance based on the recommendations of the DOT Emergency Action Guide.
3. The driver should position the apparatus to block at least one lane of the road to create a barrier between the firefighters and oncoming traffic. Ideally, the angle of the apparatus will allow the driver to view the burning vehicle from the pump operator's position and not expose the driver to oncoming traffic.
4. Secure sufficient room for an ambulance, additional apparatus, or a medevac helicopter in case additional resources are needed.

B. Safety precautions:

1. Members engaged in firefighting and rescue efforts shall wear full protective clothing and SCBA. Protective clothing shall not be removed until the possibility of reignition has been removed.
2. The presence of broken glass and other sharp objects mandates that personnel wear gloves and other safety equipment to prevent the possibility of injury even after the fire has been extinguished.

3. Stabilize the burning vehicle as soon as possible to prevent movement. This may be accomplished in a variety of ways, such as using wheel chocks, cribbing, a winch, etc.
4. If it becomes necessary to open the hood, a hatch, or other opening to make a rescue or extinguish a fire, prop open the hood or hatch to prevent accidental closing due to the failure of a spring, rod, or compressed cylinder.
5. De-energize the vehicle as soon as possible by disconnecting the battery. Take care to prevent sparks in the event flammable or combustible vapors or fluids are present.
6. If the operation takes place in darkness, illuminate the scene properly. Operating members should wear turnouts or other clothing with reflective materials to minimize the risk of being struck by other vehicles, including apparatus.
7. Also consider the impact of strobes and other warning lights on visibility. Turn them off when it is safe to do so.
8. Firefighters and rescue workers should also be aware of the hazards associated with air bags, energy-absorbing bumpers, downed or overhead power lines, hollow driveshafts, high-pressure hoses on power steering and air-conditioning systems, air-suspension systems, and the danger of exploding tires equipped with split rims, such as the wheels used on larger vehicles and some pickup trucks.
9. Establish an adequate water supply. Hose streams should be adequate for the volume of fire. Typically, the minimum size deployed should be a 1½-inch. Approach the burning vehicle from upwind whenever possible.
10. *Never* stick your head inside a vehicle while it is still burning, and *never* crawl under a vehicle to extinguish a fire.
11. A variety of combustible metals are used for engine blocks, wheels, and other components. These may react adversely with water and may require the use of specialized extinguishing agents.

C. Fuel spills:

1. A fuel tank may be punctured during an accident or may fail during a fire, thereby causing a spill or runoff. Take precautions to contain the spill and prevent environmental damage. If the fuel has not ignited, take steps to minimize the chance of ignition.
2. A variety of fuels are used in motor vehicles. Of particular concern are LNG and LPG. A leak involving either of these fuels poses special problems, as does the possibility of a fuel tank rupture due to flame impingement. It is critical that these vessels be cooled during a fire.
3. Do not remove the fuel tank cap until you are absolutely certain that there is no excess pressure in the tank.
4. Do not turn your back on a burning vehicle, and keep charged hoselines available after extinguishment in the event of reignition.
5. Prohibit smoking.
6. Prevent sparking from tools, saws, etc.
7. A fuel leak may be controlled by using lead wool, soap, wood plugs, etc., or by turning off the fuel values on an LNG or LPG tank.

IV. <u>Recommendations</u>

A. Automobile dealerships are great resources. Personnel should visit them regularly and familiarize themselves with the features of new models. Unfortunately, many of the features that make vehicles safer for the motoring public pose unique hazards for firefighters.

B. Traditionally, the motor vehicle fire has been considered by firefighters to be routine, perhaps even boring. A reexamination of our attitudes and perceptions is in order. Vehicle fires are increasingly dangerous, and firefighters should not become complacent about them.

Emergency Operations, SOP 606.01

PROGRESSIVE HOSELAYS

I. Scope

This standard applies to using fire hose to combat ground-cover and wildland fires.

II. General

A. Each engine has been equipped with (recommended) 500 feet of one-inch forestry hose, two one-inch 10/24 gpm nozzles, and two 1½-inch × 1-inch × 1-inch gated wyes to be able to extend hoselines without having to take engines off the road or other hard surfaces. Forestry hose can also be used when the terrain or weather conditions prevent the use of brush units.

B. A hoselay attack is an effective means of controlling ground-cover fires and can be used either as the entire means of suppression or as a portion of an overall suppression effort. The burning conditions, along with fuel type and topography, will determine the techniques that are used.

C. Caution: Limit hoselays to a few hundred feet and avoid lengthy hoselays near large or intensely burning fires that pose unnecessary risk to personnel.

III. Progressive Hoselays

A. One technique for extended hoselaying operations is a progressive hoselay. It consists of advancing a hoseline from an engine along the fire's edge, extinguishing the fire, then connecting another section of hose, advancing, and extinguishing more fire. This allows a fast, aggressive attack while maintaining a continuous water supply.

B. The following procedure is normally used:

1. Locate apparatus for best advantage.
2. Begin hoselay with 1¾-inch preconnect.
3. Establish an anchor point and begin extinguishment.
4. Continue extinguishment, making sure the fire line is completely out.
5. Provide additional sections of hose as needed, using wyes or reducers as needed.
6. Bring new section of hose to approximate location of the end of the initial length of hose.
7. Uncoil new section of hose.
8. Shut off nozzle.
9. Clamp hose (kink hose if clamp is not available).
10. Open nozzle to relieve pressure.
11. Attach new section of hose to the lay.
12. Make sure new hose section is free of kinks.

13. Remove clamp or kink on initial section of hose.
14. Continue extinguishment until a new section of hose is needed and repeat procedure.

Emergency Operations, SOP 606.02

WILDLAND FIRES

I. Scope

This standard was promulgated to regulate the management of incidents involving ground cover in the wildland/urban interface.

II. Definitions

A. Anchor point: A term associated with attack methods, referring to an advantageous location, usually one with a barrier to fire spread, from which to start constructing a fire line. Used to minimize the chance of being outflanked by the fire while constructing the fire line. Most anchor points originate at or near the area of origin.

B. Backfiring: Intentionally setting fire to fuels inside the control line to reduce fuel and contain a rapidly spreading fire. Used in the indirect method only.

C. Brands: Pieces of burning debris carried aloft into the convective column. May be carried outside the perimeter of the main fire by wind, causing spot fires.

D. Brush: Shrubs and stands of short, scrubby trees generally three to twenty feet in height.

E. Cat line: A fire line constructed by a bulldozer.

F. Control line: A term used for all constructed or natural fire barriers used to control a fire.

G. Crown fire: Any fire that advances from top to top of trees or brush that is more or less independent of the surface fire.

H. Fire line: The part of a control line that is scraped or dug down to mineral soil. Normally only used in wooded areas. Generally not used on grass fires.

I. Fire perimeter: The entire length of the outer edge of the fire.

J. Head of a fire: The most active part of a wildland fire.

K. Heavy fuels: Fuels of large diameter, such as logs, snags, and large tree limbs. These ignite slowly and burn slowly but produce a large amount of heat.

L. Mop-up: After the fire has been controlled, all actions required to make the fire "safe." This includes trenching, falling snags, and checking all control lines.

M. Rear of fire: The portion of a fire opposite the head. The slowest burning part of a fire.

N. Wildland/urban interface: Where native vegetation comes in contact with structures and other man-made fuels.

III. Preplans

A. Each engine company shall be responsible for developing and maintaining preplans for wildland areas located within their districts. Preplans shall identify fuel loads and types, water sources, natural fire breaks, etc.

B. Fuel Loads:

Grass: ¼ – 1 ton per acre
Medium brush: 7 – 15 tons per acre

Heavy brush: 20 – 50 tons per acre
Timber: 100 – 600 tons per acre

IV. Nomenclature of a Wildland Fire

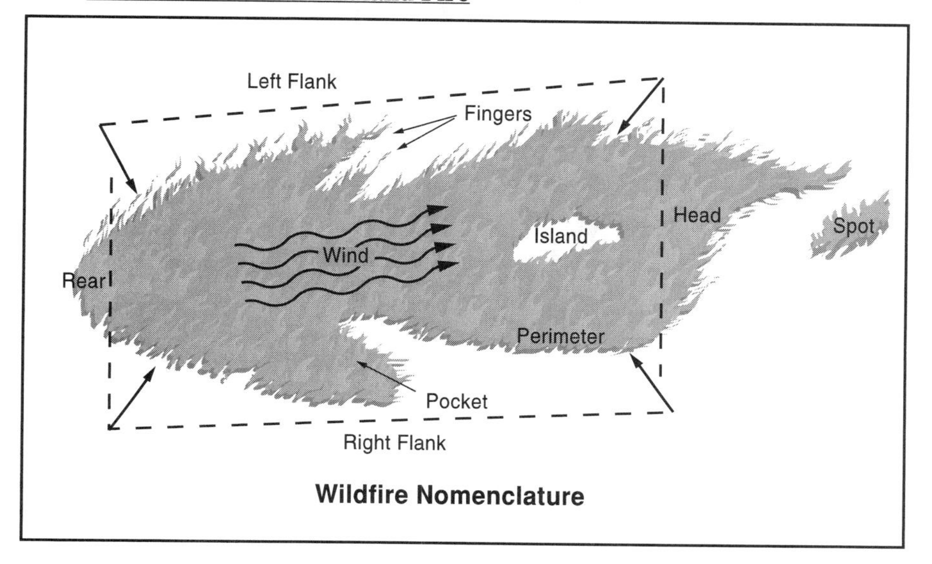

Wildfire Nomenclature

V. Response to Wildland Incidents

A. Tier I: At minimum, all reports of wildland fires shall receive an initial response of one structural engine company and one brush unit. Dispatch these units as per normal box alarm assignments.

B. Tier II: The total response on a Tier II incident shall consist of at least two structural engine companies, two brush units, a water tender, and the rehab unit.

C. Tier III: The total response on a Tier III incident shall consist of at least three structural engine companies; three brush units; three water tenders; the rehab unit; and a bulldozer from the forest service, utilities, public works, or some other source.

D. Staffing levels:

1. Structural engine: A minimum of three personnel to include but not limited to one officer, one driver, and at least one firefighter. At minimum, one crew member should be a paramedic.
2. Brush unit: Staffing will vary based on the availability of personnel and seating capacity of apparatus. A rule of thumb should be one member per door—i.e., a two-door cab equals two personnel; a four-door cab equals four personnel when available.
3. Apparatus from the same station: Stations with both a brush unit and a structural engine will be primarily responsible for responding their brush unit to wildland fires. It may be necessary to respond the engine as well to safely transport the entire crew. In such cases, the engine will *not* be considered to be the structural engine unless at least five members are able to respond from that station on the initial call.

VI. Operational Procedures

A. Direct attack: Personnel and resources work close to the fire's edge and put it out there. Best to use on small, slow moving fires with light fuels.

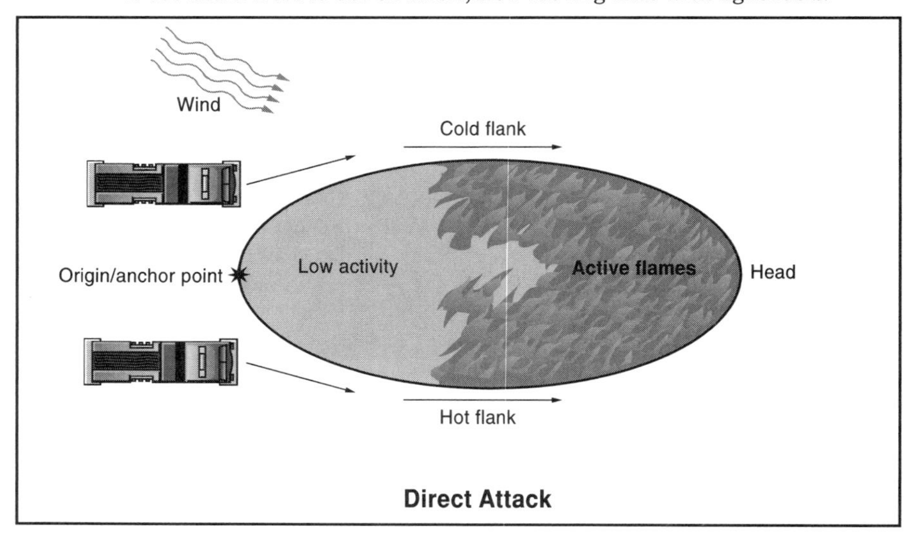

B. Indirect attack: Uses natural barriers and backfiring.

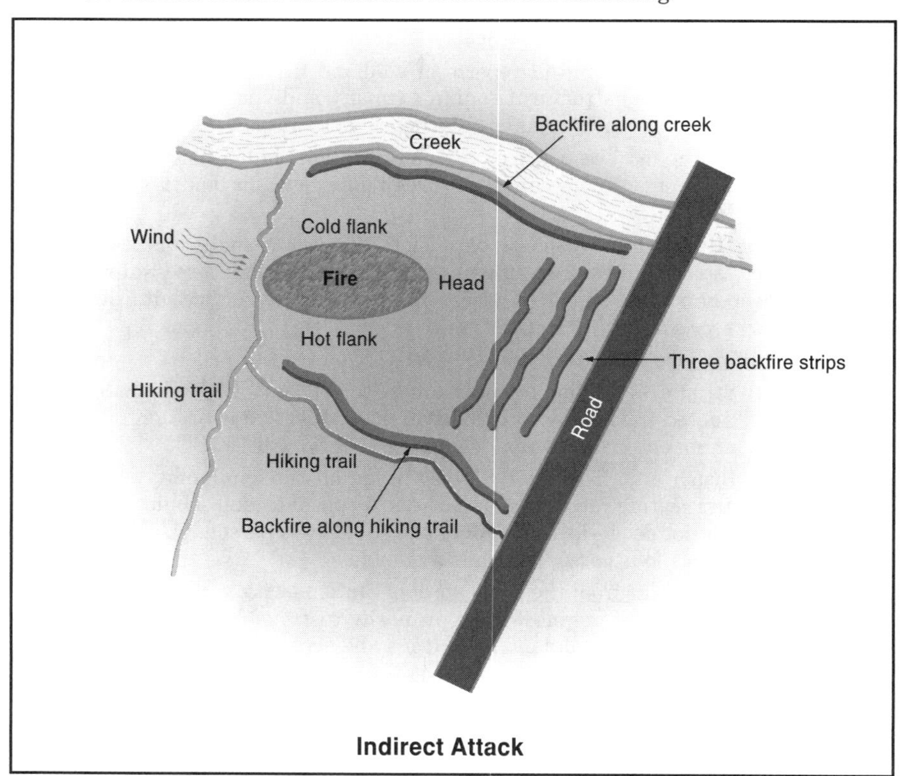

C. Parallel attack: Made by hand crews and bulldozers when intense heat or fire spread precludes direct attack. Back off five to 50 feet and parallel the flank.

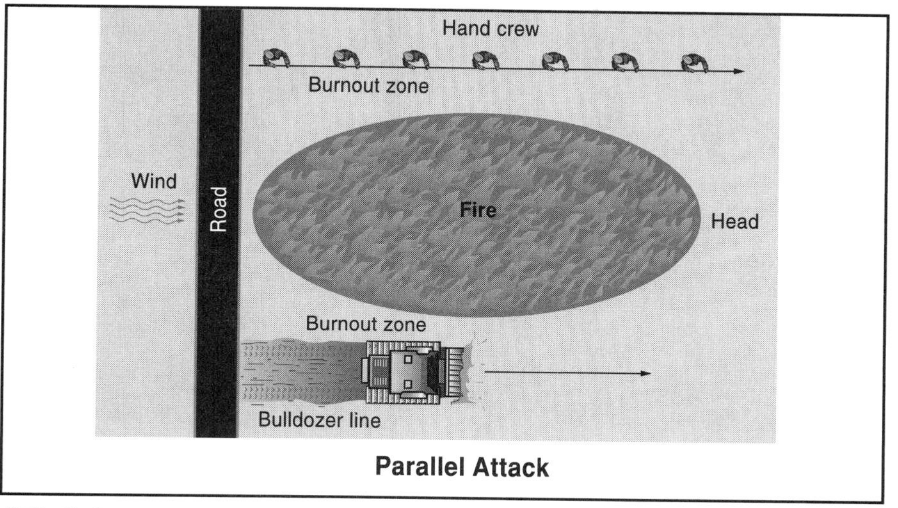

Parallel Attack

VII. <u>Safety Precautions</u>

A. The level of protective clothing to be used shall be determined by the incident commander but shall include as a minimum:

1. Department-issued NFPA-approved wildland helmet, gloves, pants and shirt, hood, and boots.
2. If *not* in department-issued wildland turnout gear, personnel shall wear structural turnouts. This includes helmet, hood, boots, turnout pants, turnout coat, and gloves.

B. Eye and respiratory protection shall be worn as conditions warrant.

C. EMS personnel shall closely monitor all personnel to prevent heat exhaustion, dehydration, etc. A rehab sector shall be established whenever appropriate.

D. The incident command system shall be used whenever more than one company operates at an incident.

E. Standard fire orders:

1. Keep informed of weather conditions.
2. Know what your fire is doing at all times.
3. Base your actions on current and expected fire behavior.
4. Plan escape routes for everyone and make them known.
5. Post a lookout when there is possible danger.
6. Be alert, keep calm, think clearly, and act decisively.
7. Maintain good communications at all times.
8. Give clear instructions and be sure they are understood.
9. Maintain control of your personnel at all times.
10. Fight the fire aggressively, but provide for safety first.
11. On grass fires, fight the fire from the burned area whenever possible.
12. Guard against getting your apparatus stuck in soft terrain. As a general rule, structural engines should not leave the roadway.

Section 700

Emergency Medical Services

GENERAL PATIENT CARE

I. Scope

This standard shall apply to all members who provide prehospital care, treatment, or transportation for the sick and injured. It was promulgated to establish guidelines for members to follow when in contact with a patient.

II. General

A. The general guiding principle of the department is to do no harm. Therefore, members shall always render whatever treatment is necessary, appropriate, and consistent with their level of training.

B. All patients shall be treated with care and respect regardless of their age, gender, race, sexual orientation, medical condition, or ability to pay for the services provided.

C. Each patient shall receive a thorough evaluation to determine his condition. Of immediate concern are:

1. Airway and respiratory maintenance.
2. Circulation.
3. Control of bleeding.
4. Stabilization of fractures.

D. The primary evaluation and stabilization shall be performed where the patient is found unless circumstances present an immediate threat to the patient or caregiver.

E. The patient shall then be stabilized and packaged for transport to a medical facility if the patient's condition warrants. The patient's condition shall be continuously monitored until he is delivered to the medical facility.

F. Transportation will be offered to every patient regardless of his ability to pay. A patient may refuse transportation, however, and will not be transported against his will unless he is incapable of making that decision due to his mental status, age, or medical condition. See SOP 701.01, Patient Transport.

G. The department will recognize a "Do not resuscitate" order if the patient's physician is present or a Directive to Physicians form has been properly executed as provided for in SOP 700.03, Terminally Ill Patients.

H. Clothing and personal property shall not be removed from the patient unless it interferes with proper treatment. The patient's modesty and right to privacy shall be strictly observed. Any search for identity shall be witnessed, preferably by a law enforcement officer. All valuables removed from the patient shall be turned over to a family member, the police, or the hospital. Valuables should be placed in a valuables envelope, the envelope shall be sealed, and a copy of the list of contents shall be turned in with the patient form.

Emergency Medical Services, SOP 700.02

DECEASED PERSONS

I. Scope

This standard contains guidelines for members to follow when a patient is obviously deceased or has been legally declared to be deceased by a competent authority. It was promulgated to:

A. Establish guidelines to assist members in determining whether resuscitation efforts should be undertaken.

B. Provide guidelines to follow when a patient is obviously deceased or has been legally declared to be deceased by a competent authority.

II. General

A. For the purposes of this standard, a competent authority shall mean a physician, a justice of the peace, or an official from the medical examiner's office. (This definition will vary by state.)

B. Members shall treat the body of a deceased patient with care and dignity regardless of the patient's age, gender, race, or sexual orientation or the circumstances surrounding the death.

C. The friends and family members of the deceased shall be treated with the utmost tact and sensitivity. Members shall be of assistance to them whenever possible.

D. On determination that a patient is deceased in accordance with the provisions of Section III, below, the body shall not be disturbed and the scene shall be preserved until the arrival of a law enforcement officer or other competent authority.

E. Members shall endeavor to return their company or unit to service as soon as possible.

F. It shall be the general policy of the department to decline transporting the body of a deceased individual unless special circumstances warrant.

III. Resuscitation Guidelines

A. The decision to begin resuscitation efforts shall be based on the department's protocol for therapy. As a general rule, resuscitation efforts shall begin unless one or more of the following conditions exist:

1. The patient is declared to be deceased by a competent authority.
2. A law enforcement official declares the incident to be a crime scene and is willing to take full responsibility for preventing the resuscitation effort.
3. An evaluation of the patient's condition reveals one or more of the following:
    a. Decapitation.
    b. Decomposition.

c. Rigor mortis.
d. Dependent lividity.
e. Visual massive trauma to the brain or heart conclusively incompatible with life.

B. Additionally, resuscitation efforts shall not be undertaken at incidents involving mass casualties, hazardous materials, or terrorist activity if so warranted by a patient's condition, a lack of resources, or if the level of risk is unacceptable.

C. Should there be any doubt as to whether or nor not to resuscitate, begin CPR immediately.

D. Once begun, do not discontinue resuscitation efforts unless:
1. CPR was initiated prior to the arrival of a fire company or ambulance and the patient's condition is obviously incompatible with life and would fit into one of the categories listed in Section III A., above.
2. The rescuers are too exhausted to continue their efforts.
3. Ordered to discontinue by a competent authority.
4. A living will or a "Do not resuscitate" order is present and satisfies the requirements of SOP 700.03, Terminally Ill Patients.

Emergency Medical Services, SOP 700.03

TERMINALLY ILL PATIENTS

I. Scope

This standard applies to the treatment of a patient who has legally executed a Directive to Physicians form. It was promulgated to:

A. Honor the wishes of a terminally ill patient who does not wish to have life-sustaining measures performed on him.

B. Provide members with guidelines to follow when dealing with a terminally ill patient who has legally executed a Directive to Physicians form.

II. General

A. The law allows individuals who are at least 18 years of age, of sound mind, and acting by their own free will to instruct their physicians not to use artificial methods to extend the natural process of dying by executing a Directive to Physicians form (living will).

B. For the Directive to Physicians form to be honored, at least one of the following conditions *must* be met:
1. The patient must be under the care of a hospice or similar organization.
2. The patient must be under the care of a nursing home or other extended health care facility and must have properly executed a "Do not resuscitate" order.
3. The apparent cause of death shall be directly attributable to the patient's illness and not result from any other cause such as choking, suicide, etc.

III. Guidelines

A. When called to aid a terminally ill patient, members shall:
1. Perform all necessary assessments to determine the level of consciousness and the degree of circulatory and respiratory function.

2. Confirm that the patient has properly executed a Directive to Physicians form and has not issued a revocation.
3. Attempt to resuscitate any patient who is pregnant or whose death appears to have resulted from other than natural causes.

B. If the patient has expired, every effort shall be made to comfort and assist the patient's family. The patient's body should then be turned over to the hospice, the police, or the medical examiner.

C. Follow normal procedures if the patient has not yet expired but is transported to a medical facility.

D. If the patient expires en route to a medical facility, no attempt should be made to resuscitate him. Curtail all fluids and medications. Leave all tubing, etc., in place and do not remove it until so ordered by the attending physician or the medical examiner.

Emergency Medical Services, SOP 700.04

MASS-CASUALTY INCIDENTS

I. Scope

This standard applies to emergency medical incidents involving five or more patients.

II. General

A. For the purposes of this standard, any emergency medical incident that involves five or more patients shall constitute a mass-casualty incident and shall be managed according to the provisions of this standard and the department's incident command system outlined in SOP 600.02, Incident Command System.

B. The first-arriving fire or rescue company at an incident with five or more patients shall notify Dispatch that a mass-casualty incident has occurred and shall institute the provisions of this standard. The incident commander shall summon additional resources as required.

C. Dispatch shall place the appropriate medical facilities on notice that a mass-casualty incident has occurred and will provide them with updates on the patients' condition and the number to be transported as more information becomes available.

III. Organization

A. At minimum, the incident commander shall designate the following sectors:

1. Triage: The immediate area where rescue operations and initial patient evaluation is being performed. Multiple triage sectors may be required depending on the magnitude of the incident.
2. Treatment: An area located a safe, convenient distance from the triage sector where victims are taken for pretransport care. Depending on the number of patients, secondary and constant triage may be necessary within this sector as well.
3. Staging: An area where personnel, ambulances, and fire apparatus report prior to being assigned to a working sector. The staging level and the number of staging areas will be determined by the magnitude of the incident.

4. Transportation: A separate area adjacent to the treatment sector where the packaged patient is assigned to an ambulance for transportation to a medical facility.

B. Each sector will be referred to by the designation indicated in Item A, above. The magnitude of the incident may require that one or more separate radio channels or frequencies be assigned to each of the sectors.

IV. Responsibilities

A. The first-arriving fire company or ambulance shall be responsible for implementing the mass-casualty procedure and for establishing command.

B. The incident commander shall be responsible for establishing the appropriate sectors and for summoning sufficient resources. All requests for additional resources must be made by the incident commander.

C. The triage sector commander should be a certified or licensed paramedic and shall:

1. Supervise the rescue, triage, and tagging of the victims.
2. Supervise the delivery of patients to the treatment sector.
3. Manage the disposition of victims who are obviously deceased as prescribed by SOP 700.02, Deceased Persons.
4. Request additional resources as required.

D. The treatment sector commander should be a certified or licensed paramedic and shall:

1. Cause a secondary and continuing triage of patients delivered to the treatment sector to prioritize care.
2. Supervise the pretransportation treatment and packaging of patients.
3. Obtain the patient's hospital destination from Medical Control and write the hospital name on the triage tag.
4. Determine the level of transportation required (ALS, BLS, bus, helicopter, etc.) and communicate this information to the transportation sector.
5. Supervise the delivery of patients to the transportation sector.
6. Request additional resources as required.

E. The staging sector commander shall:

1. Determine the level of staging required.
2. Maintain a record of the names of all emergency personnel deployed at the incident and record the amount and type of equipment managed by staging.
3. Maintain a reserve of at least one ambulance and one engine company and a sufficient number of other resources as may be required.
4. Deploy additional resources to the appropriate sector on request.
5. Request additional resources as required.

F. The transportation sector commander shall:

1. Assign and arrange patient transportation using the priorities and hospital assignments as indicated on the triage tags.
2. Maintain a record of the patients transported and their respective destinations.
3. Request additional resources as required.

V. Triage Tags

A. To assist in the prioritization of treatment and transportation, Triage Tags will be used in a mass-casualty incident.

B. Personnel assigned to rescue or field triage sectors shall evaluate each patient and attach a Triage Tag to the patient's upper or lower extremity. (The head or neck may be used as a last resort.) The following information shall be recorded on each tag:

1. Date and time.
2. Triage classification.

C. Personnel assigned to the treatment sector shall record the following information on each Triage Tag:

1. The patient's vitals.
2. Definitive therapy actions taken.
3. Injury or tourniquet areas.
4. Reclassification of triage condition as needed.

D. Personnel assigned to the transportation sector shall complete the bottom of the triage tag, remove the stub, and maintain the stub as a record of transported patients.

Emergency Medical Services, SOP 701.01

PATIENT TRANSPORT

I. Scope

This standard regulates the transportation of the sick and injured to and from hospitals and other health care facilities.

II. General

A. It is the policy of the department to offer every patient the opportunity to be transported to a hospital or other appropriate health care facility regardless of the patient's age, gender, race, medical condition, sexual orientation, or ability to pay.

B. Exception: A member may decline to transport a patient if the member reasonably believes that in doing so he will place himself or his crew in immediate danger. In this circumstance, the member shall immediately notify his supervisor of his decision to decline transportation.

C. Any patient who declines treatment or transportation to a medical facility will be required to sign a statement of refusal on the Basic EMS Report.

D. No patient shall be forcefully transported unless he is incapable of granting consent because of unconsciousness, mental status, or age. When in doubt, enlist the assistance of a law enforcement officer, child/adult protective services, or a district judge. This will vary by political jurisdiction.

E. If ordered to transport a patient by a law enforcement officer or if a patient is combative or must be restrained, request that a law enforcement officer ride with the patient in the ambulance to the medical facility as a security measure.

F. Transportation will be provided to the nearest emergency room unless otherwise directed by Medical Control, the patient's physician, or the patient. Only requests to legitimate destinations, however, will be honored.

G. Routine decisions to transport a patient to a medical facility will be based on the evaluation of the member attending the patient, direction from Medical Control, and the patient's desire to be transported. The following conditions, however, mandate transport:
   1. If the patient is less than 18 or more than 60 years old.
   2. Chest or abdominal pain.
   3. Diabetes emergencies.
   4. Seizures.
   5. Syncope episodes.
   6. Trauma cases.

H. It is the general policy of the department to decline requests for nonemergency transportation to hospitals or health care facilities located outside of the district or to local health care facilities such as nursing homes and physician's' offices. Such requests shall be referred to a private vendor.

I. Emergency and nonemergency transfers may be made between hospitals within the district when patient loads or other conditions warrant, provided that the appropriate arrangements have been made prior to the transfer.

J. It is the general policy of the department to decline transport to patients who are obviously deceased unless special circumstances warrant transporting them. See SOP 700.02, Deceased Persons.

III. Responsibilities

A. It shall be the responsibility of each member to maintain his proficiency with the department's protocols and to base his evaluation of the need to transport a patient on the protocols.

B. It shall be the responsibility of each member to offer each patient transportation to a hospital or health care facility. If the patient refuses transport, the member shall be responsible for requesting that the patient sign a statement indicating that he has declined treatment or transport. A notation shall be made on the form if the patient refuses to sign the form. Ideally, the name of a witness to the refusal should be recorded on the form.

Emergency Medical Services, SOP 701.02

EMERGENCY TRANSFERS

I. Scope

This standard regulates the transportation of the sick and injured to a hospital or other health care facility located outside of the district. It was promulgated to:

A. Establish guidelines for transporting the sick and injured to a hospital or other health care facility located outside of the district.

B. Maintain a minimum number of in-service ambulances within the department's response area by using off-duty members and reserve ambulances to make emergency transfers to hospitals and health care facilities located outside of the district.

II. General

A. It shall be the policy of the department to provide transportation of the sick and injured to a hospital or other health care facility that is located outside of the district whenever that transportation is deemed by a competent medical authority to constitute an emergency and there is no other reliable form of transportation available to the patient.

B. Such transportation shall be provided only at the request of the attending physician and after arrangements have been made at the receiving medical facility. The transport will not take place until the names of the receiving facility and the admitting physician are obtained.

C. The transferring physician shall provide all pertinent patient records, X-rays, and test results to be transferred with the patient, and those records shall be ready prior to the time of the transport.

D. The transferring facility shall provide a nurse or technician to assist in making the transfer whenever specialized life support equipment is required or if the members making the transfer deem it necessary.

III. **General Procedure**

A. The transferring facility shall provide not less than one hour's notice of the need for the transfer unless a critical patient is delivered to the facility's emergency room and the decision to make the transfer is immediate.

B. On receipt of the notice of the impending transfer, Dispatch shall issue an all-call page and request an off-duty transfer team. A transfer team requires a driver and an attendant. The attendant must be a paramedic and the driver must be approved to drive a ambulance.

C. Members who are available to make the transfer shall contact Dispatch. If a team is not secured within 10 minutes, two subsequent pages will be made at 10-minute intervals in an effort to assemble a transfer team. If a team cannot be assembled within 30 minutes, an in-service ambulance shall be dispatched to make the transfer.

D. If an off-duty transfer team is assembled, the team will use a reserve ambulance that is fully equipped to make the transfer. The transfer team must be in uniform or wear coveralls that have been assigned specifically for this purpose.

E. Full-time, career team members shall qualify for a minimum compensation of four hours of overtime or compensatory time. Their time shall start whenever they report en route to the transferring facility. However, team members shall not be released until the ambulance that has been used for the transfer has been restocked, refueled, and properly cleaned.

F. In a critical situation, an in-service ambulance may be dispatched to make the transfer, and a team may be assembled to place a reserve unit in service until the original unit is able to return to service.

Emergency Medical Services, SOP 701.03

TRANSPORT BY HELICOPTER

I. **Scope**

This standard regulates the transportation of patients by helicopter. It was promulgated to:

A. Identify those situations that dictate using a helicopter to transport patients.

B. Establish guidelines for the safe landing, loading, and liftoff of a helicopter used for medical transport purposes.

II. When to Request a Helicopter

A. A helicopter shall be called to transport a patient whenever:
   1. Instructed to do so by Medical Control.
   2. A patient's condition exceeds the capabilities of a local medical facility.
   3. Terrain or other conditions prevent access to the patient by any means other than air.
   4. A mass-casualty incident occurs and there are multiple critical patients, or the magnitude of the incident exceeds local capabilities.

B. Additionally, call a helicopter to transport a patient whenever:
   1. Traffic congestion would significantly increase the transport time to a medical facility to the detriment of the patient.
   2. The time to extricate a critically injured patient suggests transportation by an alternate means.

III. Who May Request a Helicopter

A. All requests to transport a patient by helicopter shall be channeled through Dispatch and shall not be made directly to a helicopter service or another agency.

B. The following individuals are authorized to request a helicopter:
   1. An incident commander.
   2. The first-arriving officer or paramedic prior to the establishment of command.

C. A helicopter may also be placed on standby by the individuals listed in Item III B, above, and by a crew member of a responding ambulance.

D. Dispatch shall secure authorization from Medical Control to transport by air prior to requesting the dispatch of a helicopter to an incident.

IV. General Procedures

A. Whenever conditions exist to warrant requesting a helicopter, the incident commander shall notify Dispatch to make the request.

B. The aircraft should have the ability to operate on the department's radio system. Establish radio contact with the responding aircraft as soon as possible.

C. Relay patient information to the aircraft as well as landing instructions.

D. A helicopter may also be placed on standby if there is a potential need for transportation by air but insufficient information is available to warrant an immediate liftoff. The standby mode directs the flight crew to respond to the helipad and to remain in the aircraft for further instructions. The crew can be airborne within two minutes from a standby position.
   1. The incident commander should request that Dispatch contact the helicopter service's dispatcher and that a unit be placed on standby.
   2. The helicopter service's dispatcher will confirm whether a crew is available and has been placed on standby. Dispatch should relay this information to the incident commander.
   3. If a helicopter is subsequently needed, the incident commander may request the aircraft to respond. If the aircraft is not needed, the incident commander should notify Dispatch to have the helicopter stand down. Give this notification as soon as possible due to the limited availability of aircraft.

V. **Establishing a Landing Zone**

A. The landing zone shall be as level as possible, open and away from trees and overhead wires, and free of other debris that might endanger the aircraft.

B. A clear zone of at least 100 feet in diameter shall be established and maintained. If possible, mark the four corners with smoke bombs during the day and flares at night. Place a fifth device on the upwind side of the landing area to indicate the wind direction.

C. An alternative for nighttime operations is to mark the perimeter with emergency vehicles with the warning lights operating. Use headlights and telescoping lights to illuminate the area, but do not direct them upward or otherwise interfere with the pilot's vision. Personnel shall stand by their vehicles to shut off the lights if so directed by the pilot.

D. The final decision to land shall be the pilot's. The pilot may also designate an alternate landing site if so desired.

E. A charged hoseline shall be available, if possible, whenever an aircraft lands or lifts off.

VI. **Safety Guidelines**

A. Whenever the aircraft is landing or taking off, establish a clear zone that is at least 100 feet in diameter. This zone shall be off-limits to everyone.

B. Personnel shall not approach a helicopter until signaled to do so by the pilot or a member of the flight crew.

C. Always approach a helicopter from the front, *never* from the rear! The tail rotor moves so fast that it may not be visible.

D. Walk, do not run. Always remain in view of the pilot.

E. Personnel should always wear a helmet and eye protection near the aircraft and should avoid wearing hats or other articles of loose clothing.

F. When approaching a helicopter, bend at the waist, and be aware that the main rotor blades can flex down as much as four feet. If the aircraft is on an incline, always approach from the downhill side.

G. Do not smoke within 50 feet of the aircraft.

H. Prior to and during flight, all seat belts shall be fastened and loose equipment shall be secured.

VII. **Responsibilities**

A. The pilot is always in command of his aircraft and is the final authority as to the flight and which, if any, patients are to be transported. The pilot's decisions shall be strictly followed and not questioned.

B. All personnel shall be responsible for complying with the provisions of this standard and shall commit no act that compromises the safety of the patient, the aircraft and crew, or another member of the department.

Emergency Medical Services, SOP 702.01

INFECTION CONTROL PROGRAM

I. **Scope**

This standard shall apply to all members. It was promulgated to establish a comprehensive program for preventing illness and death from occupational exposure to communicable diseases.

II. General

A. The health and welfare of each member is a universal concern. Unfortunately, exposure to communicable diseases is an occupational hazard that may occur during an emergency response or while performing a routine task. Therefore, the fire department will provide each member with the best protection available in an effort to prevent the contraction of a communicable disease.

B. To assist in this endeavor, an employee's health status shall be reviewed at least annually as specified by the provisions of SOP 406.01, Fitness for Duty.

III. Patient Contact

A. Every patient contact shall be regarded as potentially infectious and the appropriate precautions shall be taken as prescribed by SOP 702.03, Protective Clothing.

B. No patient shall be refused proper emergency medical care on the basis of a known or suspected diagnosis of a communicable disease.

C. All linen used for patient transport shall be considered contaminated and shall not be washed in a station laundry facility but will instead be exchanged at the medical facility receiving the patient.

D. All needles and other expended medical supplies shall be properly disposed of by placing them in the ambulance's biohazard container.

1. Place spent needles into the containers immediately to avoid sticking other members accidentally. Never recap a needle or stick it into a mattress, seat, or pillow.
2. Do not reuse needles.

E. Patients suspected of having an airborne communicable disease shall wear a face mask or particulate respirator during transport whenever possible, and the ambulance's ventilation system shall be turned to the maximum level.

F. Equipment that becomes contaminated as the result of patient contact shall be removed from service, tagged, and decontaminated as specified by SOP 702.02, Decontamination.

IV. Immunizations

A. As a preventive measure, all members will be offered the following immunizations and screenings on an annual basis:

1. Hepatitis B.
2. Flu shot.
3. TB screening.

B. A permanent record of all immunizations shall be maintained in the employee's permanent medical file. See SOP 406.02, Permanent Medical File.

C. Members are not required to be immunized, but those who decline must complete an Employee Immunization Refusal Form. This form shall be placed in each member's permanent medical file.

D. On the recommendation of the department's physician, a member will also be offered other immunizations, if available, after exposure to a communicable disease. All exposures shall be recorded as prescribed by SOP 406.03, Exposure Reporting.

E. Any member exposed to HIV may request periodic testing to monitor his status. The test results shall be recorded in the member's permanent medical file.

V. **General Hygiene**

A. Items contaminated by exposure to medical or hazardous substances shall either be decontaminated or properly disposed of. No contaminated waste shall be placed into station waste containers designated for normal household waste.

B. Kitchens, restrooms, laundries, and other such work areas shall not be used as decontamination areas.

C. Contaminated uniforms and other articles of clothing should be properly decontaminated and shall not be taken home to be laundered.

D. The department maintains Material Safety Data Sheets for chemicals and cleansing and disinfecting solutions. Consult the MSDS for the recommended use of individual products.

E. Stations and other work areas shall be kept clean and orderly. A clean work environment helps reduce the spread of many common communicable diseases.

F. Washing the hands with soap and water is one of the most effective tools against spreading communicable diseases. Members shall wash their hands after using the bathroom, before eating, before and after handling or preparing food, and any other time deemed appropriate.

VI. **Responsibilities**

A. The program manager for emergency medical services shall provide periodic training on infection control techniques and protective measures, as well as updates on communicable diseases as they become available.

B. The program manager for emergency medical services shall also schedule employee immunizations on an annual basis.

C. Officers shall be responsible for ensuring that all personnel under their command strictly adhere to the department's policies on infection control.

D. Each member shall be responsible for strictly adhering to the department's policies on infection control and for reporting any change in their personal health that would pose a risk to their fellow employees or patients.

Emergency Medical Services, SOP 702.02

DECONTAMINATION

I. **Scope**

This standard establishes guidelines for the decontamination of personnel and equipment. It was promulgated to:

A. Prevent the illness or death of members who come in contact with a contagious disease due to exposure to a bodily fluid or any other hazardous substance.

B. Establish procedures for the decontamination of humans, vehicles, and nondisposable equipment that has been exposed to a bodily fluid or any other hazardous substance.

II. General

A. The blood, body fluids, and tissues of all patients shall be considered potentially infectious. Portions of a vehicle, items of nondisposable equipment, areas of a caregiver's body, and articles of clothing that come in contact with these substances shall be considered contaminated.

B. Body fluids include saliva, sputum, gastric secretions, urine, feces, cerebrospinal fluid, breast milk, serosanguinous fluid, semen, or any other drainage.

C. A patient or member may also become contaminated by exposure to a hazardous material or substance.

III. Decontamination Procedure

A. Nondisposable equipment shall be properly decontaminated after every patient use. Any item that cannot readily be decontaminated shall be labeled as being contaminated and shall be quarantined.

B. In most instances, the decontamination of personnel can be accomplished by simply removing contaminated gloves and other protective devices and washing the hands and other exposed body parts with soap and water. Gloves and other disposable protective devices shall be properly disposed of as specified in Section IV, below.

C. Uniforms and articles of clothing that are contaminated shall be decontaminated in the station laundry area. The article of clothing should be removed as soon as possible. The member should shower and dress in a clean uniform. The contaminated items shall be washed in the station but not be washed with other items. Under no circumstances should a contaminated item be taken home to be cleaned due to the possibility of infecting or contaminating a family member.

D. Contaminated protective clothing shall be decontaminated by using the gear washer located at ______________________.

E. Apparatus and nondisposable equipment shall be thoroughly washed with soap and water. In some instances, it may be necessary to use a disinfectant or a bleach solution. Always follow the manufacturer's recommendations when using a cleaning product. When decontaminating items, be sure to wear gloves and other protective clothing as may be required.

F. The decontamination process shall be conducted in a maintenance area designated for this process and shall not take place in the kitchen or other living area. Exception: Restrooms may be used to shower and change clothes. Exercise caution, however, to prevent contaminating the restroom.

IV. Disposal of Contaminated Items

A. Items that cannot be properly decontaminated shall be disposed of as follows:

1. Contaminated waste items shall be placed in a garbage bag or other container, then properly sealed and labeled as contaminated waste.
2. The bags of contaminated waste shall be transported to the hospital or other designated waste handler for disposal.
3. Contaminated waste shall not be disposed of in dumpsters or other containers designated for the disposal of normal household waste.

B. Reusable bins and containers used to store contaminated waste shall be inspected, cleaned, and disinfected at least weekly. If the outside of the container becomes contaminated, decontaminate it immediately.

C. Infectious or contaminated waste shall not be stored in the kitchen or living area of a fire station prior to disposal.

V. <u>Responsibilities</u>

A. It shall be the responsibility of each member to strictly adhere to the provisions of this standard.

B. Drivers shall clean and decontaminate their assigned vehicles and equipment as soon as practical after exposure to blood, body fluids, or other hazardous or infectious material.

C. Officers shall monitor all members under their command to ensure strict compliance with this standard.

D. The program manager for emergency medical services shall conduct classes periodically on decontamination procedures and shall be responsible for updating this standard as new practices and information become available.

Emergency Medical Services, SOP 702.03

PROTECTIVE CLOTHING

I. <u>Scope</u>

This standard shall apply to any member who comes in contact with a patient. It was promulgated to prevent the contraction of communicable diseases by prescribing appropriate preventive measures.

II. <u>General</u>

A. Members who provide patient care or who are otherwise likely to be exposed to blood or other body fluids shall be provided protective emergency medical garments, face protection devices, and gloves that meet the applicable requirements of NFPA 1999, *Standard on Protective Clothing for Emergency Medical Operations.*

B. Members shall wear gloves when in contact with a patient or in the presence of bodily fluids. Patient care shall not be initiated until the gloves are donned.

C. Each member shall don a protective emergency medical garment and face protection device prior to administering any patient care whenever large discharges of bodily fluids are likely, such as during childbirth or incidents involving spurting blood.

D. Gloves and other protective devices shall be disposed of after being used, and they shall not be reused. For multiple-casualty incidents, use a new set of gloves for each patient whenever possible.

E. Contaminated garments, face protection devices, and gloves shall be placed in a red biohazard plastic bag for proper disposal after use. The contaminated biohazard bags shall be disposed of at a hospital or other designated waste handler. Under no circumstances shall contaminated items be left at the scene of an incident or disposed of in the station trash.

F. Contaminated uniforms, protective clothing, nondisposable equipment, and vehicles shall be cleaned and decontaminated as soon as practical. Decontamination shall always take place prior to being returned to service. Members shall clean and decontaminate the items in accordance with SOP 702.02, Decontamination.

G. SOP 405.01, Protective Clothing and Equipment, shall also be strictly adhered to when conditions warrant. The provisions contained within this SOP do not relieve a member of the responsibility of wearing and using the appropriate protective clothing and equipment.

H. A member who is exposed to a hazardous material or contagious disease while delivering patient care shall report his exposure as required by SOP 406.03, Exposure Reporting.

III. Responsibilities

A. The program manager for EMS shall issue a sufficient quantity of protective emergency medical garments, face protection devices, and gloves to each fire company and ambulance crew to ensure that such devices will be available for each incident to which the company or crew responds.

B. The driver of each fire apparatus and ambulance shall inspect his vehicle at the beginning of each shift and after each emergency medical incident to ensure that a sufficient quantity of protective devices are available for use by each member of the company or crew.

C. Each member shall be responsible for wearing the appropriate level of protection at an emergency medical incident.

D. Officers shall be responsible for ensuring that all persons under their command wear the appropriate level of protection at emergency medical incidents and shall immediately correct any violation that is observed.

E. Officers shall also review each member's compliance with this standard when conducting the member's performance evaluation and shall note any failures to comply.

F. Students and observers involved in the delivery of patient care shall strictly adhere to the provisions of this standard.

Section 800
Communications

Communications, SOP 800.01

GLOSSARY

The following terms are commonly used by the department in both written and oral communications:

1. Academy: A facility used to train recruits to be firefighters. May also refer to a facility or complex where in-service training is conducted.
2. Aid station: A designated location at an incident where EMS personnel treat the sick and injured. The person in charge of an aid station will normally be a paramedic and will coordinate activities with the command post. The aid station may also be divided into sectors such as Triage, Treatment, and Transportation.
3. Alarm: An incident or event that requires a response by one or more fire companies or medical units. There are several types of alarms:
   A. Automatic alarm: A request for emergency service from an alarm company or a security office on activation of a smoke or heat detector or of a fixed extinguishing system. The flow of water within a fixed system, the closure or opening of a valve, or the activation of a fire pump may also result in the transmission of an alarm signal.
   B. Box alarm: The response assignment dispatched to a reported fire in a building or structure.
   C. General alarm: An incident that requires the response of all of the department's personnel and apparatus.
   D. Greater alarm: Any alarm calling for a second alarm or heavier assignment of fire companies and personnel.
   E. Multiple alarm: A request for additional assistance at an incident to which a box alarm assignment has been previously dispatched. Multiple alarms are designated as second alarm, third alarm, etc.
   F. Still alarm: A minor incident such as an automobile or dumpster fire that requires only one company to be dispatched on the initial alarm.
4. All clear: A phrase used on completion of the primary search of a fire building indicating to all personnel that the search has been completed and that no victims were found.
5. Assignment: A predetermined designation of the units to respond to a given type of incident; the entire complement of apparatus assigned to any given incident; the assignment of any given unit.
6. Automatic aid: A programmed plan that responds the closest available company to an incident even though the closest company may be from a different political jurisdiction.

7. Back in: A term used to indicate that a company should relocate to another company's quarters for a fill-in assignment.
8. Base station: A fixed two-way radio station located either in the Dispatch office or the watch office of a fire station.
9. Buggy: The official automobile assigned to a chief officer. The term is a carryover from the time when the official vehicle was a horse-drawn buggy.
10. Call: An alarm for a fire or emergency.
11. Callback: (1) The recall of off-duty personnel back to duty for an incident or event. (2) A telephone number provided by 911 to contact a person who reports a fire.
12. Can handle: A message from a unit at the scene of an incident indicating that no further assistance will be required.
13. Catch a hydrant: An order to a responding engine company to perform a forward lay of a supply line.
14. Charge: To turn on the water and fill a hose with water and pressure.
15. Command: The radio identifier for the officer in charge of an incident. Also known as the incident commander or IC.
16. Command post: A designated location at an incident where the primary command functions are executed. The command post will be staffed by the incident commander, support personnel, and representatives from other agencies as required.
17. Controlled burn: Planned burning, allowed only by permit, conducted to remove fuel, abate a hazard, or clear a building site prior to construction.
18. Critique: A formal process following an incident and conducted by the personnel who responded so as to analyze their actions, correct deficiencies, and identify those tasks that were performed correctly.
19. Detail: The assignment of one or more personnel to temporary duty with another company or work group.
20. Detection: The act or system of discovering or locating fires.
21. Dispatch: (1) To order a fire company or medic unit to respond to a certain location, incident, or event. (2) The radio identifier for the department's emergency communications center.
22. Disregard: An order to one or more responding units that their services are not needed and that they should return to service.
23. District: A designated geographic area of service delivery normally covered by a single fire station. It may also refer to the entire area covered by a single fire department regardless of the number of stations.
24. Drill: A training session.
25. Drill tower: A multistory training structure.
26. Elapsed time: The time used to complete any assignment.
27. Emergency: A radio term used to clear the radio of all radio traffic. The term *emergency* should be followed by a specific message or set of instructions.
28. Emergency traffic: The act of clearing a radio channel of all nonessential communications.
29. En route: Indicates that an apparatus or other unit is responding to an incident.
30. Exposure: A building, vehicle, or other property that is endangered by fire in an adjacent building, a vehicle, or property.

31. False alarm: An alarm for which no fire or emergency existed or for which fire department response was unnecessary.
32. Fill in: The dispatch of another apparatus or medic unit to replace companies not available to answer their regular assignments. See also Back in.
33. Fire Alarm: Dispatch (2).
34. Fire danger: A term indicating the risk of a wildland fire due to such weather conditions as prolonged drought, high winds, low humidity, etc.
35. Fireground: The operational area at a fire.
36. Fire school: An accredited university offering regular programs in fire science. May also refer to a recruit school or training academy.
37. First due: The first company listed on an alarm assignment for a given location that is nearest in response time and travel distance.
38. First in: The first company or unit to arrive at an incident.
39. Forest Service: (1) An agency with fire control responsibility for wildland fire suppression. (2) U.S. Forest Service. An agency with fire control responsibility for wildland fire suppression in national parks, national forests, and other land owned by the federal government.
40. Front line: Apparatus and medic units normally staffed at all times.
41. Fully involved: A size-up report that indicates that the entire area of a building is so involved with heat, smoke, and flame that immediate access to the interior isn't possible until some measure of control has been achieved with hose streams.
42. Hazard: Any condition that poses a threat to property or that might result in injury or death.
43. House lights: Lights that may be controlled from the watch office or by Dispatch to illuminate a fire station when it is to respond to an alarm.
44. House phone: A private telephone that is paid for by station personnel and is used for personal calls.
45. Incident: A fire, medical call, or other emergency that requires one or more fire companies or medical units to be dispatched to render aid. See also Alarm.
46. Incident command system: A systematic plan for conducting operations during an incident. See SOP 600.02, Incident Command System.
47. Incendiary: A fire believed to have been deliberately set.
48. Incipient: A fire of minor consequence or in initial stages.
49. Initial alarm: The first notification received by the department indicating that a fire or emergency exists.
50. In service: (1) A report indicating that an apparatus or ambulance is fully functional and available to respond to an assignment. (2) A radio message indicating that an apparatus or ambulance has completed its previous assignment and is available for the next call. (3) A radio message indicating that a company or medic unit has left its quarters and will be monitoring the radio for any assignments.
51. Investigation: (1) Sending an individual, company, or unit to check for smoke, heat, steam, or other indication of fire. (2) The act of determining the cause and origin of a fire. (3) The act of determining whether or not a complaint received by the department concerning the actions of one or more of its employees was proper and within the scope of his duty.

52. Journal: A day book or record book maintained by a captain of all activities, alarms, visitors, etc.
53. Location: A specifically designated place to which fire apparatus or medical units are dispatched in answer to an alarm or request for assistance.
54. Log: A chronological record of events, such as the Dispatch Log or Incident Log.
55. Malicious false alarm: A false alarm of fire deliberately sounded to inconvenience the fire department and to cause a disturbance or excitement rather than one sounded by accident or error.
56. Message: A radio communication consisting of a contact call, response, text, and acknowledgment.
57. Move up: The movement of fire companies from their assigned stations to cover vacated stations so as to give coverage to districts stripped of normal protection.
58. Mutual aid: Two-way assistance by fire departments of two or more communities freely given under prearranged plans or contracts so that each will aid the other in time of emergency and also provide for joint or cooperative response to alarms near jurisdictional boundaries.
59. Nothing showing: A report given by the first-arriving unit at an incident indicating to Dispatch and other responding companies that no smoke, fire, or other emergency situation is apparent.
60. Operator: A fire alarm operator, dispatcher, or telecommunicator.
61. Out of service: A report indicating that an apparatus or ambulance is not available to respond to an alarm. This report should be accompanied by a message indicating the estimated length of time that the unit will be unavailable.
62. Over the air: Via radio transmission.
63. Overcome: The state of a person being incapacitated by heat, smoke, or toxic gases so as to be rendered helpless and possibly unconscious.
64. Patient: Someone who is sick or injured and requires the assistance of the department. A patient may also be referred to as a victim, citizen, customer, individual, person, man, woman, or child. A patient should *never* be referred to as a subject, perpetrator, or suspect!
65. Patrol: To travel a specified route to prevent or correct conditions that might create a hazard.
66. Permit: Official permission given in writing to allow a special activity.
67. Platoon: An organized group of firefighters who are assigned to work the same tour of duty. Also known as a shift.
68. Progress report: A periodic radio report required from an incident commander to update Dispatch on the status of an incident.
69. Quarters: The fire station to which a given company or unit is assigned.
70. Rear: The side of a building or incident directly opposite the main street front or command position.
71. Recall: To call off-duty personnel back to their stations or to a major incident.
72. Receiver: A mobile or base radio unit that allows a person to hear a radio message on a specific channel or frequency.
73. Recruit: (1) A new employee during the first 12 months of his employment. Also known as rookie. (2) The act of encouraging people to apply for employment with the department.

74. <u>Recruit school</u>: A formal training curriculum in which new employees are provided with at least the minimum number of training hours and subjects as required by law. Also known as an academy.
75. <u>Rehab</u>: This term can refer to either the actual rehab vehicle or to a designated location at an incident. The purpose of rehab is to provide rest, refreshments, and medical evaluation to working personnel.
76. <u>Rekindle</u>: An instance where, due to reignition, the department is called back to a location where the fire was thought to have been extinguished.
77. <u>Relieved</u>: (1) Used to describe a fire company that is dismissed from further duty at the scene of an emergency. (2) Used to describe the routine act of changing shifts. (3) Used to describe the temporary dismissal of an individual by a supervisor due to a pending disciplinary action.
78. <u>Repeater</u>: A radio that receives a signal from another radio and rebroadcasts the signal with greater signal strength. For example, a five-watt handheld radio doesn't have the strength to transmit to all portions of a response area. However, a repeater can receive this weaker signal and rebroadcast it with a strength of 100 watts, sufficient to cover the entire district.
79. <u>Reserve</u>: Apparatus or ambulance units not on frontline duty but available in case a frontline unit is undergoing repairs. It is also available to be staffed by off-duty personnel when necessary.
80. <u>Respond</u>: To answer an alarm in accordance with a prearranged assignment or on the instruction of the Dispatch operator. To proceed to the scene of an incident or other event.
81. <u>Responding</u>: A term indicating that orders to proceed to an alarm have been received and the apparatus or medical unit is on its way.
82. <u>Response</u>: The act of responding to an alarm. Also, the entire complement of personnel and apparatus assigned to an alarm.
83. <u>Response time</u>: An interval of time measured from the receipt of a request for emergency service until the first unit or apparatus arrives at the scene of an incident.
84. <u>Rig</u>: A fire apparatus.
85. <u>Riser</u>: A vertical water pipe used to carry water for fire protection to elevations abovegrade, such as a standpipe or sprinkler riser.
86. <u>Roster</u>: A list of fire department personnel and their duty assignments. Also, a list of apparatus and motor vehicles owned by the department.
87. <u>Run</u>: A fire or medical alarm.
88. <u>Run card</u>: The card filled out by Dispatch for each incident dispatched. Also known as the Alarm and Fire Record Card.
89. <u>Running card</u>: A card showing fire company assignments for a given location, including multiple-alarm assignments.
90. <u>Sector</u>: A specific task assignment (e.g., Staging), a geographic area (e.g., north sector), or an operational area (e.g. interior sector) of an incident that is designated and assigned by the incident commander. This is a command and control function. Sector commanders should coordinate their activities with the incident commander and use their assignment as their radio identifier. During high-rise operations, the sector designation corresponds to the floor of the building.
91. <u>Shift</u>: A working tour—e.g., 24 hours on, 48 hours off. Also refers to a group of workers on a given shift. See also Platoon.

92. Shop: The city, county, or district motor vehicle repair facility.
93. Signal: A radio message referring to the strength of a radio transmission and the listener's ability to hear and understand the message.
94. Staging: A designated location(s) at an incident where apparatus, equipment, and personnel are assembled for deployment. The person in charge of a staging sector will coordinate his activities with the incident commander and will use the term *Staging* as his radio identifier. When more than one staging area or sector is used at an incident, a geographic identifier will be used, such as *Forward Staging*.
95. Station: A building or quarters that houses on-duty personnel, apparatus, and medic units.
96. Street Index: A complete listing of all streets, roads, and highways located within the department's response district. The Street Index is organized alphabetically and numerically by block number. In addition, the index lists cross streets and major landmarks and the box number and hydrants for each intersection. The Master Street Index is housed in the dispatcher's office and is periodically updated by a fire officer assigned that responsibility.
97. Support: In incident command, those logistical functions that aid the resolution of the incident.
98. Suppression: The total work of extinguishing a fire, beginning with its discovery.
99. Tap out: A term dating from the era of telegraph fire alarm systems meaning that a fire has been extinguished.
100. Territory: A geographic area served by a single fire station or the entire area served by a department. See also District.
101. Tied up: A fire company or medical unit engaged for a period of time and unable to respond to incidents.
102. Time of arrival: The time as indicated on the radio log that the first unit arrived at an incident. Also, the time that other responding companies arrived.
103. Tour of duty: Any given on-duty period worked by an individual or group of employees.
104. Transmitter: A mobile or base radio that allows voice messages to be sent by way of a given frequency.
105. Turn-out time: The interval of time as measured from the receipt of an alarm until a fire company or medical unit reports en route or notifies Dispatch that it is responding.
106. Two-way radio: A mobile or base radio unit that allows both the transmission and receipt of audio messages.
107. Under control: A fire is sufficiently surrounded and quenched so that it no longer threatens destruction of additional property.
108. Wash down: The cleansing or removal of gasoline, diesel fuel, or other petroleum products from a roadway following a motor vehicle accident. Originally meant to wash the product down into a ditch or storm sewer. Now the product has to be collected due to environmental regulations.
109. Watch: An interval of time during which a person is assigned to a specific duty. In some jurisdictions, this duty is served at the watch desk in the watch office.
110. Watch desk: The desk in a fire station at which the various communications equipment is placed and alarms are received and recorded.
111. Watch office: An office in which the watch desk is placed.

112. Water supply: In incident command, the officer assigned to provide an adequate supply of water to meet the fire flow demand at a given incident.
113. Wildland fire: A fire involving natural groundcover such as grass, brush, and trees.
114. Working fire: A fire that requires firefighting activity on the part of most or all of the personnel assigned to the alarm.

Communications, SOP 801.01

RADIO PROCEDURES

I. Scope

This standard establishes guidelines for the use of two-way radio communications equipment. It was promulgated to promote the most efficient and effective use of the radio communication system.

II. General

A. The department operates a (insert type) ________________ radio system. The system uses (transmitter or repeater information) ________________ located at ________________. This ensures a continuous, uninterrupted source of electrical power.

B. Most of the department's radios contain ________________ channels or frequencies, which have been assigned as follows:

Police department ________________

Fire department ________________

Mutual aid ________________

Neighboring towns, districts:

____

____

____

____

____

____

____

____

____

____

____

____

C. It is the responsibility of all personnel to remain in radio contact with Dispatch while they are on duty. Therefore, they should notify Dispatch when they change location or status.

III. Restricted Activities

A. The radio system is designed for emergency communications and those activities that support the accomplishment of the department's mission. Therefore, a number of subjects are inappropriate when using the system. Common sense and good judgment should always be the user's guide when deciding the appropriateness of a message.

B. Personnel who use a two-way radio should realize that the radio does not afford the user the same level of privacy as when making a telephone call.

C. The following items are inappropriate and should never be broadcast over a two-way radio:

1. Any term that would be offensive to someone of another race or gender.
2. Profanity.
3. Any discussion of an athletic event or political contest.
4. The name of a deceased firefighter before the proper notification of family members.
5. Business of a personal nature.

IV. Channel/Frequency Assignments

This section is included to help manage a department's radio system and to instruct members on how the two-way radio communications system works. Modify it to meet local practices. A wide variety of radio systems are currently in use, ranging from CB radios to 800 mHz trunked systems. The complexity of a department's radio system will determine the length and complexity of this section. A department will typically operate in one of four ranges: low-band VHF (30-50 mHz), high-band VHF (150-170 mHz), UHF (400 mHz), or 800 mHz (which may or may not be trunked). There shouldn't be a significant difference in how the radios operate, however.

A. Frequency/channel _________ has been designated the department's primary channel. All incidents shall be dispatched on the primary channel and routine, nonemergency traffic will be conducted on this channel unless otherwise instructed by Dispatch. Therefore, all members should monitor this channel at all times.

B. Emergency operations should be conducted on a tactical frequency/channel, if available. The assignment will be made at the time of dispatch, and the responding units will be advised to move their traffic to the appropriate frequency/channel. Each incident will be assigned a separate frequency/channel whenever call volume permits.

C. At large-scale incidents, the incident commander should request Dispatch to assign a separate frequency/channel to staging, water supply, and other support operations.

D. When operating in a high-rise building or other confined space, it may be necessary to operate portable radios in a simplex mode because the density of the structure may not allow the radios to activate the repeater. Use frequency/channel _________ whenever this occurs.

E. Frequency/channel _________ has been designated as the medical control channel. Units shall use this channel to contact the hospital emergency room.

F. If other frequencies/channels have been designated to serve a specific function, include them in this section.

V. Terminology

A. Use plain speech or clear text when transmitting over a two-way radio. The department does *not* use any system of 10 codes or CB lingo. Although the department does not use numerical codes, a distinctive vocabulary of words, phrases, and terms has been developed for use in radio conversations. These terms simplify and clarify radio conversation as well as contribute to brevity. (See Part VII—Clear Text, below.)

B. The department also uses the 24-four hour clock rather than the traditional 12-hour clock. The 24-hour clock is often referred to as the military clock. All references to time used in two-way radio communications will be expressed in the 24-hour format. For example, 9:00 A.M. is expressed as 09:00 hrs (pronounced zero nine hundred hours). 9:00 P.M. is expressed as 21:00 hrs (twenty-one hundred hours).

C. Use the ICAO (International Civil Aviation Organization) phonetic alphabet to clearly identify each letter of the alphabet:

A - Alpha	H - Hotel	O - Oscar	V - Victor
B - Bravo	I - India	P - Papa	W - Whiskey
C - Charlie	J - Juliet	Q - Quebec	X - X-ray
D - Delta	K - Kilo	R - Romeo	Y - Yankee
E - Echo	L - Lima	S - Sierra	Z - Zulu
F - Foxtrot	M - Mike	T - Tango	
G - Golf	N - November	U - Uniform	

VI. Sending and Receiving Messages

A. To ensure that a radio message will be clear and understandable, the user of a two-way radio should observe the following practices:

1. Always speak in a conversational tone and at a moderate speed.
2. Speak directly into the microphone. While speaking, keep your lips within a half-inch of the microphone.
3. Remain calm. Always speak distinctly and clearly, pronouncing each word carefully.
4. Phrase your message naturally, not word for word. Avoid lengthy discussions, and be clear and to the point!
5. Use ordinary conversational strength. If surrounding noise interferes, speak louder, but do not shout.
6. Remember that a high-pitched voice transmits better than a low-pitched voice.
7. Figures, difficult words, and important messages should be repeated by the speaker as necessary. The repeated portion should be preceded by the phrase *I repeat*.

B. Message format:

1. Identify the unit or function sending the message, as well as the unit or function to whom the message is being directed.

   Example: "Engine Four to Command."
2. Wait for the unit being called to acknowledge, and then keep the message brief and to the point.

   Example: "Engine Four to Command."
   "Command to Engine Four, go ahead."
   "Engine Four to Command, the primary search is complete. We have an all clear."
3. Use procedural words and phrases whenever possible.
4. Use phonetic spelling when using words or terms that might be difficult to understand or may be spelled a variety of ways.

C. Eliminating common errors:
1. The most common error committed by a user of a two-way radio is short keying. This is caused when a radio operator attempts to transmit a message before the repeater has time to engage. This practice chops off the first part of the message.
2. To correct the problem, the user should press the transmit button on the microphone and delay his message for three to five seconds. This delay allows the repeater time to engage. An experienced radio operator can actually hear the repeater engage. Once the repeater engages, the entire message can be successfully transmitted and received.
3. An error similar to short keying results when the radio operator fails to transmit the prefix of his assigned radio identifier when reporting en route or on location when responding to an assignment. The root cause of this problem is apathy or laziness on the part of the radio operator.
4. The error is magnified when more than one unit operates out of the same station. For example, Engine 3, Truck 3, and Medic 3 are all housed at Station 3. Failure to transmit the entire radio identifier not only causes confusion, it can also result in a costly error by someone thinking he has heard a particular unit report en route or on location.

VII. Clear Text

Words and Phrases:	Application:
Affirmative	Yes.
Call by phone	Self-explanatory.
Clear	Understood.
Clear of the scene	Assignment is completed, units returning to their stations, etc.
Disregard	Cancel present assignment and return to sevice.
Emergency	Term used to gain control of radio channel to report an emergency. All other radio users will refrain from using that channel until cleared by Dispatch.
Emergency traffic only	Radio users will confine all radio transmissions to an emergency in progress or a new incident. Radio traffic that includes status information (e.g., response, conditions, location, availability) will be authorized during this period.
En route	Responding to a destination.
In quarters	Indicates a unit is in a station.
In service	On the radio, available for a call.
Loud and clear	Self-explanatory.
Major accident	A motor vehicle accident with injuries.
Minor accident	A motor vehicle accident without injuries.
Negative	No.
On location	Has arrived at the scene of an incident.
Out of service	Indicates a unit is unavailable to respond to a call.
Received	Understood.
Repeat	Self-explanatory.

Report	Provide a status update on the progress of an incident.
Resume normal traffic	Radio channel is cleared for normal use.
Return to	Self-explanatory.
Respond, responding	Indicates a unit should proceed to/is proceeding to an incident.
Stand by	Stop transmitting.
Tap out	The fire is out or is under control.
Unreadable	Radio signal is unclear. In most cases, try to add the specific trouble. Example: "Unreadable, background noise."
Weather	Self-explanatory.
What is your location?	Self-explanatory.

Communications, SOP 801.02

RADIO NUMBERS

I. Scope

This standard establishes guidelines for the assignment of radio identifiers to apparatus and personnel equipped with two-way radios.

II. General

A. A number of systems exist for the identification of fire and rescue apparatus. Many systems are based on a regional mutual aid agreement and are intended to reduce confusion when apparatus from multiple jurisdictions are operating at the same incident. A department should identify in this section the type of system to be followed. The system used by the North Central Texas Fire Chiefs Association Mutual Aid Network is included as an example in the table *Radio Identification Prefixes,* below.

B. It is useful when apparatus can be identified by function, jurisdiction, and station assignment. This may be accomplished by using either a number or a prefix followed by a number.

1. Function: Apparatus can be identified with a prefix that designates its principal function, such as engine, truck, or ambulance. It is also possible to use a number series, e.g., 100s for chief officers, 200s for engines, 300s for truck companies, etc.
2. Jurisdiction: It is very common to use a number identification to indicate the jurisdiction. For example, the departments in Northern Virginia (NOVA) use the following designation: 100s = Alexandria, 200s = Arlington County, 300s = Metropolitan Airports Authority (Dulles and National), 400s = Fairfax County, and 500s = Prince William County.
3. Station identifier: This may be accomplished by using the station number as one or more digits in a series of numbers.
   a. In the first example, Unit 101 would be the battalion chief from Station One.
   b. In the NOVA example, Truck 101 would be a ladder truck from the City of Alexandria's Station One.
   c. If a station has multiple units of the same type, a variety of systems can be used. For example: Engine 201A or Engine 201-1.

Another example would be Engines 1021 and 1022. The first two digits would be the department number, the third digit is the station number, and the fourth digit is the unit number.

III. **Fire Department Radio Identifiers:**

Radio number:	Assigned to:

IV. **Portables**

A. Most fire apparatus and ambulances have been assigned one or more handheld two-way radios. Use the following designations whenever transmitting:

1. A captain will use the letter A following the apparatus identifier:
   Example: "Engine 163A to Engine 163, charge the green line."
2. Drivers will use the designation B, and firefighters will use C:
   Example: "Engine 163B to Engine 163C, I am charging your supply line."
3. On medic units, the driver or a captain will use the A designation, and the attendant will use the B designation.

B. All administrative and fire prevention personnel have also been assigned portable radios. When an inspector or other individual uses his portable radio, he will simply use his assigned radio number.

Radio Identification Prefixes

The North Central Texas Fire Chiefs Association has adopted the following radio identifiers:

Resource Type	Radio ID	Computer ID	Function
Admin. chiefs	Chief	C	Chief, deputy chief, etc.
Airport crash truck	Crash	CR	Air crash rescue
Brush rig or booster	Brush	BR	Wildland firefighting units
Engine company	Engine	E	Engines, pumpers, etc., 500+ gpm
Haz mat	Haz mat	HM	Hazardous materials teams
Inspectors	Marshal	I	Inspectors, investigators, etc.
Medic or ambulance	Medic	M	ALS/ BLS EMS transport and support
Rescue	Rescue	R	Rescue: high angle, swiftwater, auto extrication, etc.
Shift commander	Battalion	B	Battalion, district, incident commander
Staff and support	Admin.	AD	Fire Administrative Staff Members
Truck company	Truck	T	Ladders, quints, snorkels, towers
Utility, light, or air supply	Utility	U	Lights, SCBA air, rehab
Water tender	Water tender	W	Water transport of $\geq$ 1,000 gallons

Communications, SOP 802.01

RADIO MAINTENANCE

I. **Scope**

This standard regulates the maintenance and repair of two-way radios and communications equipment.

II. **General**

A. Two-way radios, pagers, base stations, and station alerting systems are all vital components of the department's emergency notification and communications system. It is imperative that all equipment function properly on demand.

B. Members who have been issued a pager, two-way radio, or other communications device are responsible to maintain their assigned equipment in proper working order.

C. Captains (company officers) are responsible for the proper operation of all communications equipment assigned to their command.

III. **Repair Procedure**

A. Whenever a two-way radio or other piece of communications equipment is in need of repair, the proper service agency shall be notified. If the repair facility is open, the item shall be taken to the shop for repair before the facility closes for the day.

B. Station alerting systems and other essential equipment are critical to emergency operations and must be repaired immediately. Therefore, the on-call technician should be notified and asked to respond as soon as possible.

C. Not all equipment is of such a critical nature that it has to be repaired immediately. If a replacement item is available or if an alternative method of communications is available, the repairs may be postponed until the repair facility is open for business. For example, if the two-way radio in an apparatus doesn't operate properly, it will usually be possible for the apparatus to use a portable radio until the mobile unit can be repaired.

D. All communications equipment owned by the department shall be maintained and repaired by ______________________________. The department's repair facility is located at ______________________.

E. Pagers, telephones, and other communications equipment leased from a private vendor shall be returned to the proper vendor for repair in accordance with the lease agreement.

F. Disposable items such as batteries may be obtained by contacting the communications officer.

G. All repairs shall be reported to the communications officer. The communications officer shall keep a record of all the repairs made to the department's equipment and shall include a summary of the repair activity in his monthly report to the fire chief.

Communications, SOP 803.01

DISPATCHING ALARMS

I. **Scope**

This standard establishes guidelines for dispatch and communications personnel to follow when receiving requests for service and dispatching emergency units. It was promulgated to ensure that appropriate response assignments are dispatched.

II. **Receiving an Alarm**

A. Prior to dispatching fire companies to an alarm, the dispatch operator must obtain sufficient information to properly dispatch the alarm. At minimum, this information includes the traditional variables of who, what, when, and where. Other clarification may be required.

B. The dispatch operator should ask the person reporting an incident the following questions:

1. What is your name and telephone number?
2. What is the nature of your emergency?
3. What is the exact address or location of the emergency?
4. As the dispatch operator records this information, he will note the correct time that the call was received.

C. If the request for service is received via 911, the dispatch operator should compare the information given by the caller with the information displayed on his computer screen. The operator should note any discrepancies and question the caller further to make sure that the information given is correct.

D. *All* of this information should be recorded on an Alarm and Fire Record Card. The operator should stamp the card with the automatic clock to record the time and date that the alarm was received.

E. The dispatch operator should then enter the address of the incident into his computer to obtain the correct box number, cross street, map page number, and other pertinent information. If the computer is inoperable, this information may be located manually in the Master Street Index.

III. **Dispatching the Proper Units**

A. Every incident shall be broadcast on the department's primary radio channel and any other appropriate channels. The following information shall be broadcast and repeated at least once:

1. The type or nature of the incident
2. The location (address).
3. The nearest cross street or landmark.
4. The units to respond.
5. The radio subfleet assigned to the incident.
6. The map or box number.
7. The time of dispatch.

Example: "Structure fire, 217 West McKinney at Cedar. Engines 3, 5, 2, and 4; Truck 3; and Medic 2 respond on Tac 1. Map 33B. Repeat. 16:30 hours."

B. This information is to be recorded on the Alarm and Fire Record Card, which shall also contain a chronology of events pertinent to the incident. The dispatch operator shall review the record on conclusion of each incident and shall correct any discrepancies.

Example: An incident was dispatched as a structure fire at 123 Elm Street. On arrival, the fire was actually found to be located across the street at 124 Elm Street.

C. It is not uncommon to have more than one incident working at a time. To better manage the communications function, each incident should be assigned to a separate subfleet or channel whenever possible. Therefore, the operator shall assign a specific channel to each incident and shall identify the response channel at the time of dispatch.

D. Whenever an ambulance is dispatched to another jurisdiction that contracts with the department for EMS, Dispatch shall notify the appropriate fire department and request that its first responders be dispatched to the incident.

E. The response assignment depends on the location of the incident and the type of emergency reported to Dispatch.

F. As a matter of practice, *always* dispatch the units that are closest to the incident. The closest apparatus to any given location may be determined in several ways. For example:

   1. The dispatch operator should be familiar with the location of major streets, target hazards, and landmarks.
   2. The status and location of all apparatus should be constantly monitored and recorded. This will allow the operator to determine the availability of a particular unit whenever an alarm is dispatched.

      Example: A call comes in for a dumpster fire on the college campus. This is in Station 3's district. However, the operator knows that Engine 3 is at the city shop getting fuel and that Engine 4 is at Administration. Engine 4 is closer to the incident than Engine 3. Therefore, Engine 4 should be dispatched.
   3. Look up the box number listed in the Street Index. The box number lists the proper order of response to the location. The box number is a four-digit number. This number lists four fire stations in the order of their proximity to the location in question.

      Example: Box 4253. Station Four would be the closest Station. Station Two is the second closest, and so forth. If all companies are in service in their respective stations, the companies should be dispatched in the order indicated by the box number.
   4. If the fourth digit is a zero, dispatch the water tender in addition to the appropriate assignment for the type of occupancy.

      Example: Box 4250. The response to a single-family dwelling would be Engines 4, 2, and 5; Truck 3; Medic 2; and the water tender.

G. The actual response assignment is determined by the nature of the emergency and the information currently available to Dispatch.

   Example: Only one ambulance is normally dispatched to an MVA. If, however, the MVA is reported by a police officer who indicates that the accident involves a school bus with multiple injuries, the number of medic units dispatched would depend on the number of injuries that the officer reports.

IV. <u>Fires</u>

A. Automatic alarms:

   1. (Actual practice will vary by jurisdiction, but the following is strongly recommended.) On receipt of an automatic alarm, the operator shall dispatch one engine company, nonemergency. After the engine company

has been dispatched, the operator shall attempt to verify the validity of the alarm. This may be done in several ways.

   a. The receipt of a 911 call reporting a fire or other emergency at or near the address of the automatic alarm.
   b. Confirmation of a valid alarm by a fire or police official at the scene.
   c. Telephoning the location of the incident and verification by a person at the address of the alarm that there is indeed an emergency.

2. If an automatic alarm is determined to be a valid emergency, the response shall be upgraded to a full first-alarm assignment, and all companies shall be instructed to respond emergency.
3. If the alarm is discovered by Dispatch to be accidental or invalid, the responding company shall be notified and shall proceed to the location to make a report.
4. (Note: Local practice may vary.) The *only* time that a responding unit shall disregard an automatic alarm is on confirmation by a law enforcement officer or fire official who is actually at the location of the incident and reports that there is no emergency.

B. Hazardous materials emergencies:
   1. Dispatch one engine company for spills and transportation accidents unless the caller provides information that would dictate a greater emergency.
   2. Dispatch a full first-alarm assignment to incidents that occur within a structure.
   3. Additional assistance is available from ______________________________
   ____
   4. During a working haz-mat incident, Dispatch shall provide the incident commander with a weather update every 30 minutes, or more frequently if so requested. This information should include the temperature and the wind speed and direction.

C. Multiple alarms:
   1. Large fires that exceed the capabilities of the first-alarm assignment.
   2. Second alarm (depends on local resources; the following is exemplary):
      a. Companies from your department remaining in service.
      b. Emergency callback of off-duty members; page the fire chief, deputy chief, and fire marshal.
      c. Anytown's truck company.
      d. Disasterville VFD to Station 2.
      e. Middletown VFD to Station 3.
      f. Put private ambulance company on standby to answer EMS calls.
   3. Third alarm:
      a. Respond any staffed reserve companies to fire.
      b. Disasterville and Middletown to the fire.
      c. Request a truck company from Rio Vista.
      d. Timpson, Tenaha, Bobo, and Blair to fill in.

4. Fourth alarm:
    a. Timpson, Tenaha, Bobo, and Blair to the fire.
    b. Request a truck company from Eastown.
    c. Sodom, Gomorrah, and Jerusalem to fill in.
5. Fifth alarm:
    a. Sodom, Gomorrah, and Jerusalem to fire.
    b. Request a truck company from Iola.
    c. Request Lisbon, Madrid, London, and Paris to fill in.

D. Public service calls:
1. Events that are not immediately life-threatening and require the response of only one engine company, nonemergency. (Local practice may vary.)

   Examples: Courtesy unlocks, washdowns, smoke investigations, etc.
2. Dispatch a courtesy unlock of a vehicle only if there is a child or pet locked inside the vehicle, the motor is running, or the incident poses a danger to life or property. Otherwise, advise the caller to contact a locksmith.

E. Still alarms:
1. Minor fires that do not involve or threaten a structure are known as still alarms and require that only one engine company be dispatched.

   Examples: Automobile fires; MVAs without injuries; dumpster, trash, and rubbish fires.
2. If this type of fire threatens to spread immediately to a building or is threatening human life, then dispatch a full first-alarm assignment.

F. Structure fires:
1. Residential assignment: For fires in one- and two-family dwellings, three engine companies, one truck company, and one ambulance.
2. Light commercial assignment: Small free-standing businesses such as convenience stores, fast-food restaurants, etc. Small to average occupant loads and light fuel loads. Dispatch three engine companies, one truck company, and one ambulance.
3. Commercial assignment: strip malls, malls, churches, schools, apartment complexes, hospitals, nursing homes, and manufacturing facilities with large occupant and fuel loads. Dispatch four engine companies, one truck company, and one ambulance.
4. In an area without fire hydrants, also dispatch the water tender. These areas are indicated in the street index by a zero appearing as the fourth digit of the box number.
5. On confirmation of a working fire in a structure, Dispatch shall contract the electric and gas utilities and ask them to respond to the incident. The police department shall be notified for traffic control and asked to respond to the scene on the same tac channel. Also advise the water department and the fire marshal's office.

G. Wildland fires:
1. Fires involving grass, brush, trees, or other groundcover.
2. Refer to SOP 606.02, Wildland Fires.
    a. Tier I Response: Small incidents requiring the dispatch of one engine company and one brush unit.

b. Tier II Response: Large incidents that threaten structures. Two engine companies, two brush companies, the water tender, and the rehab.

c. Tier III Response: Large incidents that involve structures and require mutual aid. Three engines, three brush units, three tenders, the rehab, and a bulldozer from utilities, the county or the state.

d. Additional tenders are available from ______________ ____.

H. Time checks:

For all working fires, Dispatch shall contact the incident commander and provide an update of the elapsed time of the incident every 10 minutes for the first hour of the incident and every 30 minutes thereafter.

V. **Medical Calls**

A. All emergency medical calls should receive a minimum response of one engine company and one ambulance.

B. Major accidents that require extrication equipment will also require the dispatch of an engine with a rescue tool.

C. For medical calls outside the city limits, send the closest ambulance and request the appropriate fire department to respond as well.

D. Notify the appropriate law enforcement agency when the department responds to the following types of medical calls:

1. Major accidents.
2. Calls involving a weapon—i.e., gunshot wounds, suicide attempts, cuttings, stabbings, etc.
3. Any call that involves a domestic disturbance.
4. Fights, beatings, etc.
5. Anytime the dispatch operator has information that indicates the need for a police officer.

E. Time checks:

During the duration of an EMS incident, Dispatch shall provide the units at the scene with an update of the elapsed time at 10-minute intervals.

F. Multiple-casualty incidents: See 700.04, Mass-Casualty Incidents.

G. Whenever all of the ambulances are committed to incidents, the private ambulance company shall be placed on standby for any subsequent call.

VI. **Dispatching Support Services**

A. Law enforcement: Anytime a fire or EMS unit is dispatched to an incident requiring the units to respond with lights and sirens operating, the appropriate law enforcement agency shall be notified and informed of the location and type of incident.

B. Utilities: Notify the appropriate electric and gas utilities of all working structure fires. Notify other utilities on request of Command.

C. Other types and forms of assistance may be requested by an incident commander. The dispatch operator should be able to locate the type of service requested by consulting the resource lists maintained in the Emergency Management Plan or in one of the several telephone lists maintained in the dispatch office.

VII. Fill Ins

Staffing reserves:

1. During emergency callbacks, off-duty personnel shall staff reserve engine companies and medic units.
2. Reserve companies shall notify Dispatch and Command of their status.
3. Anytime the number of in-service companies drops to only one engine company, off-duty personnel shall be paged to return to duty.

VIII. Assigning Incident Numbers

A. An incident number shall be assigned to every response made by the department. The incident number shall be assigned in chronological order by time and date and shall be unique to each incident.

B. This number shall be recorded in the Master Incident Log along with the location, time, date, and type of incident.

C. The incident number shall be composed of six digits, with the first two digits corresponding to the last two digits of the year of dispatch.

Example: Incident 980001 was the first alarm dispatched in calendar year 1998.

D. Multiple incident numbers shall not be assigned to any incident regardless of the number of personnel and apparatus that may ultimately respond.

Communications, SOP 803.02

DAILY RADIO TESTS

I. Scope

This standard outlines the procedures for testing two-way radios and station alerting systems. It was promulgated to:

A. Ensure that all two-way radios and station alerting systems function properly on demand.

B. Establish a procedure to make announcements to all on-duty personnel via the two-way radio system.

II. Daily Radio Test

A. Dispatch shall conduct a test of the station radio alerting system each morning at shift change. As part of the test, announcements that are appropriate for the entire shift shall be broadcast over the primary channel.

B. The test format is as follows:

1. After paging each station individually, the dispatch operator shall broadcast the following messages:
   a. "WXYZ 123, your fire department testing. Testing 1, 2, 3, 4, 5. 5, 4, 3, 2, 1."
   b. "Attention all stations," then give the day's announcements. These should include:
      (1) Staffing is/is not complete.
      (2) Announcement of scheduled training, street closings, hydrants out-of-service, etc.
2. On conclusion of the announcements, Dispatch shall make the following statement:

   "WXYZ 123, your fire department clear," then give the time.

III. <u>Testing Individual Radios</u>

A. Test a radio whenever there is concern that it is functioning improperly. The individual checking the radio should transmit a request to Dispatch for a radio check over the primary channel.

Example: "Engine 4A to Dispatch, radio check."

B. Dispatch should then report the strength and clarity of the signal.

C. Take radios that are not functioning properly to the communications shop for repair as soon as possible.

D. The station captain (company officer) should contact Dispatch if concern arises about whether or not a station's alerting system is functioning properly. Dispatch shall conduct a test of the station's alerting system to make this determination.

Communications, SOP 803.03

DISPATCHING THE EMERGENCY MANAGEMENT COORDINATOR

I. <u>Scope</u>

This standard specifies the circumstances under which the emergency management coordinator shall be notified. It was promulgated to provide notification to the emergency management coordinator that existing conditions might require activation of the city, county, or district Emergency Operations Center (EOC) or implementation of any of the provisions of the local emergency management plan.

II. <u>General</u>

A. The law requires that each political subdivision within the state prepare an emergency management plan. This assigns the responsibility for the preparation and maintenance of the disaster preparedness plan to the emergency management coordinator.

B. A copy of the emergency management plan is on file in the dispatch office for use in the event of a disaster.

C. (Insert governing board) ______________________________ has assigned the responsibility for emergency management to the fire chief.

D. In the event of a disaster, the fire chief/emergency management coordinator is responsible for activating the EOC and providing advice and assistance to (insert appropriate officials and governing board) __________________ ____.

E. The emergency management coordinator or his designee shall be notified whenever any of the following conditions exist:

1. The issuance of a severe weather statement by the National Weather Service.
2. Confirmation of a tornado or other weather-related emergency by a weather official, law enforcement officer, or other responsible party.
3. On request of the police department in the event of civil unrest.
4. On the request of the incident commander at a hazardous materials accident or spill.
5. On the request of the incident commander at a multiple-casualty incident.
6. Anytime that special circumstances warrant notification.

Communications, SOP 803.04

PAGING

I. Scope

This standard establishes guidelines for use of the department's paging systems.

II. General

A. Specify the type of paging system operated. There are essentially two types of pagers: radio and telephone. Radio pagers are more limited in reach but offer immediate notification and may include a voice function. Telephone pagers may not be reliable as a primary means of notification but are useful for contacting members. Telephone pagers are typically either digital (only provides the receiver a number to return the call) or alphanumeric (provides a textual message).

B. If the department leases alphanumeric or digital telphone pagers from a private vendor and issues them to its members, indicate so in this section alone with the name of the vendor, if appropriate.

C. Telephone paging companies provide a wide range of service, from local to global coverage. Specify the coverage provided: ______________________

____

D. Include a list of pager assignments in this document. This will allow dispatchers and others to reach every member of the department when necessary.

Pagers have been assigned as follows:

Name:	Pager number:	Range:
Fire chief	__________	Statewide.
Deputy chief	__________	Statewide.
__________	__________	__________
__________	__________	__________
__________	__________	__________
__________	__________	__________
__________	__________	__________
__________	__________	__________
__________	__________	__________
__________	__________	__________

E. Include instructions for using a telephone pager in this document.

III. Radio Paging System

A. Radio pagers have been issued to __________________ and will be used as the principal method of notifying members of an incident.

B. The radio paging system is owned/leased/shared by the department and uses a frequency of __________. The system is actuated by the dispatcher and may be used to contact individual members, select groups, or the entire membership.

C. Include instructions for the proper use of the pager by the membership.

D. It may be useful to include a list of the individual groups and a discussion of how to activate the individual, group, and all-call functions if this document is to be used by the dispatchers.

Communications, SOP 804.01

ENHANCED 911

I. <u>Scope</u>

This standard establishes guidelines for the receipt of calls for emergency assistance by way of an enhanced 911 telecommunications system.

II. <u>General</u>

The enhanced 911 system is the most technologically advanced way of reporting and receiving emergency telephone calls currently available to public safety agencies.

III. <u>Answering 911 Calls</u>

A. The emergency number for fire and emergency medical services in this response area is 911.

B. Whenever anyone located within this area dials 911, the call is forwarded via selected routing to the communications center located at ____________. If the call has been improperly routed, the operator shall transfer the call to the appropriate emergency response agency.

C. The operator should confirm that the call is for an actual emergency. Persons who do not have an emergency shall be instructed to call the appropriate administrative number in either the fire or police department.

D. Through the enhanced 911 system, operators have been provided Automatic Number Identification (ANI) and Automatic Location Identification (ALI). These systems display the caller's telephone number, name, and address on the operator's computer scene. The operator shall verify that the information being given by the caller corresponds to the information displayed on the screen. Any discrepancies should be noted on a 911 B form.

E. In the event that the caller is unable to speak or otherwise communicate with the operator, the call shall be considered to be a valid emergency and assistance will be dispatched to the address displayed on the operator's screen.

F. All incoming 911 calls are recorded.

G. The operator shall *never* assume that multiple calls for an emergency are for the same incident. The operator should verify that the calls are for the same incident before disconnecting the caller.

H. If the emergency is for a fire or EMS incident, the operator shall record the following information on the Alarm and Fire Record Card:

1. The location of the emergency. (A rural mail route and box number are not sufficient as an address.)
2. The type of emergency.
3. The names of the persons reporting the emergency.
4. The reportee's telephone number.
5. The nearest crossroad or landmark.
6. Any other information pertinent to the incident, such as geographical identifiers, the numbers of patients, their condition, etc.

Communications, SOP 804.02

EMERGENCY RING-DOWN CIRCUITS

I. Scope

This standard establishes guidelines for use of the emergency telephones located at each fire station, intended for use by citizens to report an emergency whenever a fire station is unattended.

II. Obtaining Assistance Via the Emergency Ring-Down Circuit

A. While the department encourages citizens in need of emergency assistance to call 911, it may not always be possible to do so. Occasionally, help may be more rapidly summoned by traveling to the nearest fire station.

B. Unfortunately, fire stations are often unattended. This may be the case in the event that the apparatus and crew have responded to an incident, are conducting company inspections, are attending a training session, or any number of other legitimate reasons.

C. To provide citizens with a means of reporting an emergency in the event that a fire station is unattended, the department has installed special emergency telephones that operate on a ring-down circuit at each of the fire stations.

D. The emergency telephones are accessible to the public 24 hours a day. A sign explaining the proper method for using an emergency telephone has been placed next to each one.

E. To use the phone, all a caller must do is open the weathertight box and lift the receiver off the hook. This will cause the emergency ring-down circuit telephone to ring in the communications center until a dispatch operator answers the phone. Once the operator answers the call, the person in need of assistance will be able to speak directly to the operator.

Communications, SOP 804.03

CELLULAR TELEPHONES

I. Scope

This standard regulates the use of cellular telephones. It was promulgated to maximize the efficiency of the department's emergency communications system by incorporating cellular telephones.

II. General

A. Cellular telephones have been installed in all in-service ambulances, engines, and trucks. Additionally, most administrative and fire prevention personnel have been furnished with cellular telephones.

B. The telephones are leased from ______________________________ and will generally operate within the region. Administrative personnel will be provided with service statewide.

C. Cellular phones are to be used strictly for conducting the department's business. They are intended to be a backup or adjunct to the two-way radio system.

III. Guidelines for Use

A. Cellular telephones should be used:

1. At the discretion of the paramedics to enhance patient care.
2. During response, to obtain better directions to the site of an incident.
3. To begin patient assessment or to provide prearrival instructions to persons at the scene.
4. Whenever security or confidentiality concerns warrant use of a phone rather than a two-way radio.
5. To make direct contact with a hospital or physician.
6. To contact a supervisor.
7. To contact family members of patients, business owners, or other individuals with an interest in an incident.
8. To contact support services or other individuals or groups for information or assistance.
9. Transmission of EKGs to an emergency room.

B. Persons who are operating from a landline or other cellular telephone may also contact individuals or apparatus equipped with cellular telephones whenever it may be desirable to do so.

IV. **List of Department Cellular Telephone Numbers**

____

____

____

____

____

____

Communications, SOP 804.04

DEPARTMENT TELEPHONE SYSTEM

I. **Scope**

This standard establishes guidelines for use of the department's telephone system.

II. **General**

A. This jurisdiction's phone service is provided by ____________________, and the jurisdiction lies within the ____________________ area code. The department's telephone system functions as a component of the city's system. The individual phone sets are made by (insert vendor) ____________________.

B. The main business number for the department is ____________________.

1. To reach another number within the system, ____________________ ____.
2. To reach a local number that is outside the jurisdiction's switch, ______ ____.
3. To place a long-distance call, ____________________ ____.

C. All telephones shall be answered promptly and politely. Business lines shall be answered "Fire department" or by giving the station number—e.g., "Station Three."

D. Members shall never be rude or discourteous to anyone over the telephone.

E. The department's telephones are to be used for official business only. Members shall not be permitted to use fire department phones to conduct personal business or part-time work activities.

F. Personal calls shall *not* disrupt work or training activities. Out of respect for coworkers, personal calls shall be limited to the evening hours, and conversations shall be limited to five minutes.

G. Long-distance calls shall be limited to department business. Personal long-distance calls, if necessary, shall be limited to personal emergencies.

H. Each member of the department is required to have a telephone at his residence and to provide the department with his telephone number. At no time, however, shall a member's home telephone number be furnished to anyone other than another member of the department.

I. No telephone shall be installed in or removed from a fire station or other facility without the consent of the fire chief. Personal phones, whether portable or cellular, shall not be used while on duty in a fire station or other department facility.

III. Voice Mail

A. Several members of the department have been provided with voice mail. The voice mail system allows callers to leave messages when the number they call is busy or unattended.

Section 900

Fire Prevention

Fire Prevention, SOP 900.01

WEAPONS

I. Scope

This standard shall apply to all members of the department who are authorized by statute to carry a firearm.

II. General

A. No member of the department may carry or possess a firearm while on duty unless he has been so authorized by the fire chief and has completed the training required by (insert enabling statute or agency) ____________ ____.

B. A member who has been authorized to carry or possess a firearm while on duty shall keep his weapon accessible at all times. Firearms, however, may only be carried in the following limited circumstances:

1. While making or assisting in an arrest.
2. While serving or assisting in the serving of a warrant.
3. While actively conducting an investigation and the investigator is involved in an activity that potentially places him in harm's way. This shall include interviewing or interrogating a suspect.
4. During the course of rendering aid or assistance to a fellow peace officer.
5. While conducting a life-safety inspection or code check in a bar or other establishment that serves alcohol.
6. While issuing a citation.
7. While participating in firearms training.
8. While performing a special assignment that warrants carrying a weapon.

C. A member shall prominently display his badge or identification whenever he is armed. As a general rule, weapons should be kept concealed whenever it is possible or practical to do so.

D. A member shall be prohibited from carrying a firearm in the following circumstances:

1. While performing routine administrative activities.
2. While conducting a routine inspection or participating in a public education program.
3. While participating in a training class, evolution, or program.
4. While off duty. (This is a recommendation. Consult the local enabling statute for guidance on this issue.)

5. While attending a conference, seminar, or training program outside the district.
6. While participating as a member of a fire suppression or emergency medical crew.
7. While under the influence of alcohol or other controlled substance.

E. Only those weapons that have been authorized by the department for use by its members may be carried. No member is authorized to carry a specific weapon unless he has qualified with the weapon. An authorization form shall be kept on file with the department for each weapon carried by a member.

F. A member may only carry one firearm and shall not carry a backup or reserve weapon.

G. Firearms shall be kept clean at all times and shall be maintained in proper working order.

III. <u>Restrictions on Use</u>

A. No member shall draw or exhibit his weapon except for general maintenance, storage, or authorized training unless circumstances create a strong suspicion that it may be necessary to lawfully use it.

B. No member shall point or direct a firearm at a person unless circumstances create a strong suspicion that it may be necessary to lawfully use it.

C. No member shall surrender his weapon to any person who is not a police officer for any reason except for maintenance.

D. Firearms shall be secured on and off duty in such a manner as to ensure that no unauthorized person will have access to or gain control of weapons.

E. No member shall discharge a firearm at or from a moving vehicle.

F. No member shall discharge a firearm if in so doing it creates a substantial risk of harm to an innocent bystander.

G. No member shall discharge a firearm for the purpose of a warning shot.

H. No member shall discharge a firearm for the purpose of killing an animal unless:

1. It is necessary to kill the animal to prevent harm to the member or to others.
2. The animal is so sick or badly injured that humanity requires its relief from further suffering and it is not possible to dispose of the animal in any other manner.

IV. <u>Responsibilities</u>

A. A member who has been authorized to carry a firearm shall be responsible for complying with the provisions of this standard and shall maintain the certifications and licenses required by statute to lawfully carry a firearm.

B. The fire marshal shall be responsible for:

1. Ensuring that all persons authorized to carry firearms:
   a. Maintain their proficiency with the firearm that they have been authorized to carry.
   b. Complete the required weapons training.
   c. Maintain the required certifications and licenses required by statute to lawfully carry a weapon.
2. Maintaining a record of all firearms authorized for use by the members of the department.

Fire Prevention, SOP 900.02

THE USE OF DEADLY FORCE

I. Scope

This standard applies to those members authorized by the department to perform law enforcement duties and who are authorized by statute to carry a weapon. It was promulgated to establish guidelines for determining when it is appropriate for a member to use deadly force.

II. General

A. The use of deadly force is one of the most serious acts that a law enforcement officer performs and is contrary to the reverence for human life normally exhibited by officers who often risk their own lives to save others.

B. It is the philosophy of this department to accomplish its mission as efficiently as possible with the highest regard for the human dignity of all persons. The members of the department shall not rely on the use of physical force unless all other means have been exhausted. The use of physical force shall be restricted to those circumstances authorized by law.

C. The amount and degree of force that may be used to effect lawful objectives are determined by the circumstances and include but are not limited to:

1. The nature of the offense.
2. The behavior of the suspect against whom force is to be used.
3. The feasibility or availability of alternative actions.

D. Members shall weigh the circumstances of each individual case and employ only the minimal amount of force necessary and reasonable to control a situation, effect an arrest, overcome resistance to arrest, or defend himself or others from harm.

E. When the use of force is necessary, the degree employed shall be in proportion to the amount of resistance employed by the person or the immediate threat the person poses to the member or to others.

F. The use of force is progressive in nature and may be in the form of advice, warnings, persuasion, verbal encounters, physical contact, use of a nonlethal weapon, or deadly force.

G. Nondeadly force may be used in instances where the member reasonably believes it is immediately necessary to take action to:

1. Preserve the peace, prevent the commission of an offense, or prevent serious bodily injury;
2. Make a lawful arrest or search, overcome resistance to a lawful arrest, or prevent an escape from custody; or
3. Defend against unlawful violence to a person or property.

H. Physical force involves actual physical bodily contact with a person and forcibly subduing that individual until resistance is overcome. Physical force excludes the use of weapons or objects that might be used as weapons. In no instance shall a member use physical force in excess of that which is reasonable and necessary to lawfully and properly neutralize an assault or overcome resistance by a person being taken into custody.

I. Deadly force is force that is intended or known by the member to cause, or in the manner of its use or intended use is capable of causing, death or serious injury. Deadly force is permitted when:

1. The action is in defense of human life, including the member's own life;
2. In defense of any person in immediate danger of serious physical harm and all other means have been reasonably exhausted or are inappropriate for the circumstances; or
3. It is necessary to effect the capture of a suspect or prevent the escape of an arrestee when there is reasonable belief that he poses an immediate threat of serious bodily injury to the officer or another person and all other means have been reasonably exhausted or would be inappropriate for the circumstances.

III. **Reports on the Use of Force**

A. A member shall file a written report with the fire marshal whenever he has taken an action that results in or is alleged to have resulted in injury or death to another person. The report shall be completed prior to the completion of his tour of duty.

B. All reports shall be reviewed by the internal affairs investigator (or other designated officer—e.g. the deputy chief) and the results of the investigation shall be forwarded to the fire chief.

C. With the exception of a firearms training session, a member shall file a written report with the fire marshal anytime he discharges his weapon, whether it misses or causes injury or death to a person or animal.

Fire Prevention, SOP 901.01

COMPANY INSPECTIONS

I. **Scope**

This standard establishes guidelines for fire suppression and emergency medical personnel assigned to conduct fire prevention inspections in their first-due areas.

II. **General**

A. It is the goal of the department to prevent fires and to save lives and property. In pursuit of this goal, it shall be department policy to inspect all nonresidential properties at least twice each calendar year.

B. A secondary goal of the inspection process is to allow all members to become familiar with the buildings within their first-due response areas. The information obtained during an inspection will be of assistance in the preparation of a tactical survey for the occupancy.

C. The fire marshal shall manage the company inspection program.

1. The fire marshal shall develop and maintain a current, up-to-date master inspection list to include all buildings, businesses, and occupancies except one- and two-family dwellings.
2. The master inspection list shall be subdivided as required. Separate lists will be developed for each fire station's first-due response area and for the occupancies that will be inspected by the fire marshal's office. These lists shall contain the following information:
   a. The name and address of the building, occupancy, or business.
   b. The occupancy code.
   c. The inspection file number.

d. The first-due district.
e. The size and height of the occupancy.

2. A permanent inspection file will be maintained for each building, occupancy, and/or business. A copy of each inspection report will be placed in the file each time the occupancy is inspected.
3. A monthly report that lists the number of inspections conducted during the prior month by each fire and EMS company and the fire marshal's office will be prepared by the fire marshal. The report shall indicate the percentage of assigned inspections completed by each company and member of the fire marshal's office.
4. The fire marshal shall be responsible for the correction of all violations that are discovered. This shall be accomplished by reinspections and/or the issuance of citations.
5. Instruction in the proper techniques for conducting fire prevention inspections shall be provided to each member by the fire marshal's office.
6. The fire marshal shall meet with members periodically to review problems and to answer questions that may arise.

III. Conducting the Inspection

A. Company officers shall be responsible for inspecting all of the buildings and occupancies assigned to their company by the fire marshal.

B. Each company officer shall receive his monthly inspection assignments by the 25th day of each month. These lists shall designate the inspection of each occupancy at least twice each calendar year. The lists shall also be distributed so that each shift inspects all of the occupancies within the first-due response area at least once every 18 months.

C. There is no charge for the initial inspection. If one or more violations are found during an initial inspection, it will be necessary to conduct a reinspection. A fee is charged for each reinspection and is listed on the Reinspection Fee Schedule. The amount of the fee is based on the size of the occupancy, and it increases with each subsequent reinspection.

D. A total of two reinspections will be conducted before a citation is issued. Citations may only be issued by a member of the fire marshal's office.

E. The success of the company inspection program depends on the goodwill and voluntary compliance of the owner or manager of each business. A member's politeness and professionalism will go a long way toward making the inspection program a success.

F. On entering the occupancy, the company officer shall ask to speak to the manager or supervisor. The officer shall introduce and identify himself to the responsible party and explain the purpose of his visit. He should ask the responsible party to accompany the inspection team. This will allow the team to explain the principles of fire prevention as well as point out and correct hazards.

G. For smaller occupancies, it may be desirable to divide the company into two or more inspection teams to minimize the distraction to customers or workers. This will also allow more inspections to be done in less time in multiple occupancies. However, at no time shall any inspection be conducted by fewer than two members.

H. The exterior of the building or occupancy should also be surveyed to determine the location of doors, windows, utility meters and shutoffs, construction features, etc. Note also the two closest fire hydrants or water sources.

I. Many hazards can be corrected immediately on discovery. Note minor violations on the Inspection Report Form. No reinspection will be required if all of the violations can be corrected during the inspection.

J. All areas of the building or occupancy shall be inspected in an orderly and systematic manner.

K. The results of each inspection or reinspection shall be accurately recorded on the Inspection Report Form. Forward the pink and yellow copies to the fire marshal and give the white copy to the occupant. The yellow copy will be returned to the company officer if a reinspection is necessary.

L. If a problem arises during an inspection that cannot be solved by the company officer, a fire inspector or the fire marshal shall be requested to respond to the location to solve the problem in a timely, orderly fashion.

M. Some violations are serious enough to require that a citation be issued immediately. Such violations include locked or obstructed exits, exceeding the posted occupancy load, or any violation that constitutes an immediate threat to health or safety.

N. On discovery of such a violation, the inspection team shall request that a member of the fire marshal's office be dispatched to the location.

O. If a problem is discovered that involves an automatic fire sprinkler system, standpipe, or other suppression/detection system, the inspection team shall notify the fire protection engineer (FPE) or other appropriate official such as the assistant fire marshal or senior inspector. This person shall be requested to respond to assist in correcting the problem.

P. After completion of the inspection, the Inspection Report Form should be signed by the owner or other responsible party. Give the white copy to the responsible party. Provide an explanation concerning any hazards noted and the manner in which they should be corrected. If a reinspection is necessary, explain the reinspection procedure in detail to minimize any misunderstanding.

Q. If no hazards are found, check the box on the Inspection Report Form stating that the premises were found to be reasonably fire safe at the time of the inspection.

R. Thank the responsible party for his assistance prior to leaving.

S. In the event that the company officer is unable to complete an assigned inspection, he shall forward a written explanation to the fire marshal explaining the reason.

IV. **Reinspections**

A. When violations are found during a routine inspection, it will be necessary to conduct a reinspection after the owner or manager has had a reasonable period of time to correct the violations.

B. The first reinspection will normally be scheduled 15 days after the original inspection. A second reinspection, if necessary, will normally be scheduled 10 days after the first reinspection. If the violations are not corrected by the time of the second reinspection, a member of the fire marshal's office will be assigned to conduct a final reinspection.

C. The fire marshal's office will inspect the occupancy five days after the second reinspection and will issue a citation if the violation has not been corrected.

D. Reinspections will be assigned by the fire marshal and should be conducted in a similar manner as a routine inspection. Only those items found to be deficient during the original routine inspection should be checked.

E. Serious violations discovered during a reinspection that were not discovered or that did not exist at the time of the routine inspection should be recorded, and the occupant shall be asked to correct those deficiencies as well.

Fire Prevention, SOP 901.02

DETERMINATION OF CAUSE AND ORIGIN

I. Scope

This standard establishes that every fire and explosion incident in the district be investigated by the department to determine its origin and cause. It was promulgated to:

A. Establish guidelines for the safe and systematic investigation or analysis of fire and explosion incidents.

B. Assist in the compilation of data.

C. Prevent similar events from occurring in the future.

II. General Policies

A. It shall be the policy of the department to determine the origin and cause of fires and explosions that occur within the response area.

B. Investigations and analysis of fires and explosions shall be conducted in accordance with the recommendations contained in NFPA 921, *Guide for Fire and Explosion Investigations.*

C. Fire cause will be classified in one of the following categories:

1. Accidental: Not the result of a deliberate human act.
2. Natural: Without human intervention—e.g., lightning.
3. Incendiary (arson): Deliberately set.
4. Undetermined: The cause cannot be proved.

III. Responsibilities

A. The incident commander shall:

1. Ensure that an accurate determination of the origin and cause of a fire or explosion has been made prior to conducting overhaul operations or concluding operations and returning to service.
2. Request that an investigator be dispatched to an incident in the following circumstances:
   a. The incident commander is unable to accurately determine the origin and cause of a fire or explosion.
   b. There is evidence to suggest that the fire was deliberately set.
   c. A civilian or firefighter was injured or killed.
   d. An automatic fire detection or suppression system failed to activate properly.
3. Maintain custody of the scene and shall not allow evidence to be disturbed until the investigator authorizes the incident commander to

remove debris or other materials from the incident site or authorizes suppression personnel to leave the scene.

B. Members shall:
   1. Observe the conditions on arrival and report anything out of the ordinary to the incident commander or investigator.
   2. Preserve any evidence that they discover and not disturb any debris or other materials until so ordered by the incident commander.

C. The investigator shall:
   1. Conduct a thorough investigation of each incident to which he is dispatched and make a determination as to origin and cause in accordance with Section II C, above.
   2. Prepare investigation reports and maintain case files as required.
   3. Make arrests, serve warrants, and testify in court as required.

Appendix

Contents

1. Employee Leave Request Form, *259*
2. Daily Activity Report, *260*
3. Driver's Daily Apparatus Checklist, *261*
4. Station Supply Requisition Form, *262*
5. Report of Equipment Lost or Destroyed, *263*
6. Library Book Check-Out Log, *264*
7. Audiovisual Materials Check-Out Form, *265*
8. Complaint Form, *266*
9. Request for Travel/Training Form, *267*
10. Annual Employee Performance Review Form, *268*
11. Basic Incident Report, *272*
12. Basic EMS Report, *274*
13. Basic Casualty Report, *276*
14. EMS Charge Sheet, *278*
15. Hazardous Materials Exposure Form, *279*
16. Notification of Accident Report, *280*
17. On-the-Job Injury Supervisor's Report, *282*
18. Supplemental Report of Injury, *284*
19. Designation of Duty Status Form, *286*
20. Authorization to Ride an Apparatus Release, *287*
21. Inspection Report Form, *288*
22. Flashover/Backdraft Report, *290*
23. Rehab Log, *291*
24. Rehab Unit Checklist, *292*
25. Equipment Service Request Form, *293*
26. Small-Vehicle Weekly Checklist, *294*
27. Annual Fire Pump Service Test Form, *295*
28. Preventive Maintenance Worksheet, *296*
29. Red Tag Out-of-Service Card, *298*
30. Master Hose Record Card, *299*

31. Monthly SCBA Inspection Form, *300*
32. Ground Ladder Record Form, *301*
33. Tactical Worksheet, *302*
34. Minimum Company Standards Evaluation Form, *304*
35. Tactical Survey Form, *307*
36. Directive to Physicians, *310*
37. Triage Tag, *312*
38. Employee Immunization Refusal Form, *316*
39. Alarm and Fire Record Card, *317*
40. Reinspection Fee Schedule, *318*

EMPLOYEE LEAVE REQUEST FORM

Name:	Date:

Rank/Shift/Station:

Indicate Type of Leave Requested:

_____ Comp time _____ Military _____ Sick leave/injury

_____ Holiday _____ Shift swap _____ Vacation

_____ Other (explain): ____

Reason for Request:

Amount of Leave Requested:

Leave to Commence On:

Date: _________ Time: __________ hrs.

Return to Duty On:

Date: _________ Time: __________ hrs.

Accumulated Leave Balances:

Holiday: __________ Vacation: __________ Sick leave: __________ Comp time: __________

Signatures:

Employee:	Date:
Company officer:	Date:
Staffing officer:	Date:
Fire chief:	Date:

DAILY ACTIVITY REPORT

Date:	Station:	Shift:

Employee No.	Name	Rank	Duty Status	Company Assignment	On Duty	Off Duty

Record of Alarms:

Alarm No.	Time	Location	Type of Incident	Units Responding

Activities:

____

____

____

____

____

Officer's Signature:

DRIVER'S DAILY APPARATUS CHECKLIST

I.D. No.	Date:
Odometer Reading:	Hourmeter Reading:
Onan I.D.:	Onan Hourmeter:

Items to Be Checked:	Pass	Fail
Aerial device operates properly	______	______
Auxiliary equipment in place and operational	______	______
Back-up alarm operates properly	______	______
Battery condition	______	______
Brake fluid level	______	______
Brakes operate properly (service and parking)	______	______
Coolant level	______	______
Doors and compartments open and close properly	______	______
Fuel level	______	______
Lights / turn signals / clearance lights / horn	______	______
Oil level	______	______
Pump operates properly	______	______
Radios / cellular phones	______	______
Spare tire condition	______	______
Steering, suspension, drive components	______	______
Tire pressure: LF ___ RF ___ LR ___ RR ___	______	______
Transmission level	______	______
Check underside of vehicle and floor for fluid leaks	______	______
Upholstery condition	______	______
Warning devices	______	______
Washed	______	______
Wheels, lugbolts	______	______
Windshield wipers / washer fluid level	______	______

Comments / missing equipment / repairs needed:

Officer's signature:	Driver's signature:

STATION SUPPLY REQUISITION FORM

Item	Quantity	Description	Stock No.	Price	Total

Person Making Request:	Date:
Location:	Shift:

REPORT OF EQUIPMENT LOST OR DESTROYED

Person Making Report:	Unit/Shift:	Date:

List of Lost/Destroyed Items:

List each item and include the asset number and replacement cost. Use the back of the form if necessary.

1)		
2)		
3)		
4)		
5)		

6) How and when was the item(s) lost or destroyed?

7) When was the item(s) discovered to be lost or destroyed?

8) Was the loss/destruction an act of negligence? If so, explain.

9) How was the item replaced?

10) Describe corrective actions taken:

Report Reviewed by:

☐ Company Officer:	Date:
☐ Equipment Officer:	Date:
☐ Fire Chief:	Date:

LIBRARY BOOK CHECK-OUT LOG

Name of Borrower	Title of Book	Date Issued	Due Date	Date Returned

AUDIOVISUAL MATERIALS CHECK-OUT FORM

Title of Item	Date Issued	Due Date	Name of Borrower	Date Returned

COMPLAINT FORM

I, ________________________________, wish to make a complaint against ________________________________. I am alleging that the member ________________________________ based on the following facts:

Date occurred:	Time occurred:

Explain:

It is an offense if a person, with intent to deceive and with knowledge of a statement's meaning, makes a false statement under oath.

Signature of complainant	Date	Telephone number

This instrument was acknowledged before me

on the __________ day of (month) ________________________, (year) ______.

____
NOTARY PUBLIC IN AND FOR THE STATE OF ______________________.

Supervisor taking complaint	Date	Time
Supervisory investigating	Date completed	Date/time complainant notified

Note: Attach all statements, investigative reports, and recommendations.

Complaint Declared:

☐ Unfounded ☐ Exonerated ☐ Not sustained ☐ Sustained

REQUEST FOR TRAVEL/TRAINING FORM

Date:	**Officer Making Request:**

School/Seminar:

Dates: From to

Program Synopsis: (Use back if necessary)

Benefit to the Department: (Use back if necessary)

Personnel to Attend: (Use back if necessary)

1.	4.
2.	5.
3.	6.

Costs:

Registration	$
Accommodations	$
Meals	$
Transportation	$
Other (Itemize)	$
Total	$

Approved	**Officer**	**Initials**
☐ Yes ☐ No	Supervisor	
☐ Yes ☐ No	Shift Staffing Officer Overtime required? ☐ Yes ☐ No	
☐ Yes ☐ No	Shift/Division Commander	
☐ Yes ☐ No	Fire Chief	

ANNUAL EMPLOYEE PERFORMANCE REVIEW FORM

NAME: ______________________________ DATE: ______________

POSITION/TITLE: ______________________________ LAST REVIEW: ______________

PART I: PERFORMANCE MEASURES

Instructions:

The rating supervisor shall assign a numerical rating from 1 to 5 for each job factor listed below. Scores of 1, 2, or 5 must be accompanied by a comment. No comments shall be added to this form after the member signs the form. A copy of the completed, signed form shall be provided to the member and a copy shall be placed in the member's permanent personnel file.

Rating/Description

1 Unsatisfactory. Performance is significantly below minimum expectations and must be improved during the next rating period if the member is to continue to be employed. The rating supervisor shall provide the member with a written plan of action to improve his performance.

2 Below expectation. Performance is deficient and below that which is expected but is not totally unsatisfactory.

3 Meets expectations.

4 Exceeds minimum expectations. Above average.

5 Outstanding. Performance is clearly superior to other employees in the same or similar job and significantly exceeds expectations.

PART II: PERFORMANCE RATING

Item Performance Factor **Rating**

1. Drives assigned apparatus in a safe, efficient manner. Driving record is clear.

   Comments: ______________________________

2. Knowledge of streets, hydrants, target hazards, standpipes, sprinkler connections, and routes in assigned first-due area.

   Comments:____________________

3. Station duties and maintenance on assigned apparatus and equipment.

   Comments:____________________

4. Compliance with SOPs and safety standards.

   Comments:____________________

5. Performance in training evolutions and at fire and EMS incidents.

   Comments:____________________

6. Seeks to improve knowledge and skills through independent study, etc. Participation in wellness program.

   Comments:____________________

7. Personal appearance, hygiene, uniform, etc.

   Comments:____________________

8. Absenteeism/tardiness.

   Comments:____________________

9. Relationship with coworkers, citizens, supervisors, etc.

   Comments:____________________

10. Assists in training coworkers; assumes responsibility in the absence of supervisor.

   Comments:____________________

TOTAL SCORE:

ANNUAL EMPLOYEE PERFORMANCE REVIEW FORM, *continued*

PART III: SUMMARY

Note: Use back of form if more space is required.

A. Specific actions to be taken during next rating period:

1. ____________________
2. ____________________
3. ____________________
4. ____________________

B. Has your supervisor explained your performance rating to you?

☐ Yes ☐ No

C. Do you agree with your supervisor's assessment of your performance?

☐ Yes ☐ No

D. Supervisor's summary comments:

____

____

E. Employee's comments:

____

____

F. Comments of next higher level of supervisor:

____

____

PART IV: APPROVALS

Member's signature ______________________ Date __________

Evaluating supervisor's signature ______________________ Date __________

Shift commander's signature ______________________ Date __________

Fire chief's signature ______________________ Date __________

BASIC INCIDENT REPORT

A	FD ID	Incident No.	Index No.	Mo./Day/ Year	Alarm Time	Time on Scene	Time Last Unit Clear
B	Location/Address	City/Town	Zip Code	Property No.			
C	Occupant Name (Last, First, MI)	Telephone No.	Room or Apt.				
D	Owner Name (Last, First, MI)	Address	Telephone No.				
E	Method of Alarm to Fire Department	Type of Situation Found					
F	Type of Action Taken	District	Shift	No. of Alarms	Outside Fire Service Assistance		
G	General Property Use	Specific Property Use	County	Census Tract			
H	No. of Fire Suppression		No. of Emergency Medical Services		No. of Other Fire Service		
	Apparatus	Personnel	Apparatus	Personnel	Apparatus	Personnel	
I	No. of Injuries		No. of Fatalities				
	Fire Service	Non-Fire Service	Fire Service	Non-Fire Service			
J	Condition of Fire on Arrival of First Unit	Area of Fire Origin					
K	Equipment Involved in Ignition	Year	Brand Name	Model	Serial No.		
L	Form of Heat of Ignition	Material First Ignited					
		Form	Type				
M	Ignition Factor	Method of Extinguishment					
N	Property Loss	Number of Acres Burned					

O	Type of Construction	No. of Stories	Level of Origin			
P	Structure Status	No. of Occupants at Time of Incident				
Q	Material Contributing to Fire Growth Form	Type				
R	Factor Contributing to Flame Travel	Avenue of Smoke Travel				
S	Detector Type	Detector Power Supply				
T	Detector Performance	Reason for Detector Failure				
U	Type of Automatic Sprinkler System	Coverage of Sprinkler System				
V	Sprinkler System Performance	No. of Sprinkler Heads Operated	Reason for Sprinkler System Failure			
W	Extent of Flame Damage	Extent of Smoke Damage	Extent of Extinguishing Agent Damage			
X	Mobile Property Type	Year	Make	Model	Serial/VIN No.	License No.
Y	Member Making Report	Date	Officer in Charge (Name, Position, Assignment)	Date		
Z	Remarks:					

BASIC EMS REPORT

HA	FD ID	Incident No.	Casualty No.	Date	Alarm Time	Time on Scene	Time Unit Clear
HB	Location/Address	City/Town	Zip Code	Property No.			
HC	Method of Alarm to Fire Department	Type of Situation Found					
HD	Type of Action Taken	District	Shift	No. of Alarms	Outside Fire Service Assistance		
HE	General Property Use	Specific Property Use	County	Census Tract			
HF	Casualty Name (Last, First, MI)	Date Injury Occurred	Time				
HG	Home Address	City	State	Zip	Telephone No.		
HH	Affiliation	D.O.B.	Age	Gender	Race	National Origin ☐ Hispanic?	
HI	Case Severity	Primary Apparent Symptom	Primary Part of Body				
HJ	Secondary Apparent Symptom	Secondary Part of Body					
HK	Casualty Type by Situation Found	Final Disposition of Casualty					

	Time of Reading	Blood Pressure		Pulse		Respiration	
		Systolic	Diastolic	Rate	Character	Rate	Character
HL							
HL							
HL							

	Lungs		Skin		Pupils		
	Sound	Location	Color	Temperature	Size	Reactivity	Position
HM							

HN	Patient Status	Patient Behavior

HO	Prehospital Care Provided (1)	Prehospital Care Provided (2)	Prehospital Care Provided (3)	Prehospital Care Provided (4)

	Time		Cardiac Condition/Assessment	Drug/Fluid	Rate	Route
HP	1					
HP	2					
HP	3					

HQ	Time EKG Transmitted	Medical Facility EKG Transmitted to	Receiving Hospital Representative Signature

HR	Type of Unit Handling Medical Emergency	Responder Medical Training Level

HS	Member Making Report	Date	Officer in Charge (Name, Position, Assignment)	Date

HT	Remarks:

BASIC CASUALTY REPORT

GA	FD ID	Incident No.	Index No.	Casualty No.	Date Injury Occurred	Time
GB	Casualty Name (Last, First, MI)			Date Injury Reported		Time
GC	Affiliation	D.O.B.	Age	Gender	Race	National Origin ☐ Hispanic?
GD	Home Address	City	State	Zip		Telephone No.
GE	Case Severity	Primary Apparent Symptom		Primary Part of Body		
GF	Secondary Apparent Symptom			Secondary Part of Body		
GG	Casualty Type by Situation Found			Final Disposition of Casualty		
GH	Familiarity With Incident Area	Condition of Person Prior to Incident		Activity at Time of Injury		
GI	Location in Relation to Point of Origin			Location at Time of Injury		
GJ	Cause of Injury or Accident			Factors Preventing Escape		

GK	Regular Fire Service Work Assignment		Physical Condition at Time of Injury	
GL	Status Before Alarm		Fire Service Activity	
GM	Where Injury or Accident Occurred		Cause of Injury or Accident	
	Protective Equipment			
GN	Type	Use		Performance
GO	Manufacturer	Model	Serial or Lot No.	National Standard
GN	Type	Use		Performance
GO	Manufacturer	Model	Serial or Lot No.	National Standard
GN	Type	Use		Performance
GO	Manufacturer	Model	Serial or Lot No.	National Standard
GP	Member Making Report	Date	Officer in Charge (Name, Position, Assignment)	Date
GQ	Remarks:			

EMS CHARGE SHEET

Date:	Incident No.	Patient's Last Name

DISPOSABLES

	Item Used	No. Used	Size
31.	Airway, naso, any size		
33.	Airway, oral, any size		
14.	Airway, PTL		
46.	Bandage, triangular		
15.	Bandaging/splinting		
104.	Bandaid, any size		
20.	Blanket, emer., yellow		
51.	Blanket, reuseable		
35.	Catheter, butterfly		
28.	Cath., intraosseous		
32.	Catheter, intravenous		
27.	Catheter, mediport		
11.	Cervical coll., any size		
29.	Cold pack		
80.	Cotton-tipped appl.		
88.	CPR administered		
44.	Dextrostix, any type		
112.	Diatek cover		
10.	Dressing, multitrauma		
49.	Easy cap, nelcor		
7.	Electrodes, any size		
13.	Emesis basin		
43.	Endo-lock		
30.	E.T. tube, any size		
89.	Excessive wt., > 300 lbs.		
19.	Eye shield		
24.	Eye pad		
23.	Gauze, 4 X 4, single		
23.	Gauze, 4 X 4, 10 pack		
47.	Gauze, petroleum		
103.	Gel, defibrillation		
75.	Gloves, exam		
3.	Gloves, sterile		
45.	Gown, disposable		
4.	Head bed		
116.	Head bed (head on)		
68.	Hot pack		
110.	Jelly, KY lubrication		
54.	Kling, 4"		
100.	Knife, emer. cricoid		
94.	Lancet, any type		
21.	Liner, rico suction		
60.	Mask, dual, shield		
105.	Mask, surgical		
82.	MAST, infl.		
57.	Monitor, defib., pace		
18.	Nebulizer, updraft		
106.	Needle, any size		
26.	OB kit		
8.	Oxygen		
9.	Oxygen delivery, any		
101.	Oxygen, tubing		
58.	Pads, pacer		
90.	Pt carried down stairs		
95.	Pillow, disposable		
59.	Resuscitator, disp.		
77.	Sheet, disposable		
96.	Sheet, reusable		
34.	Sheet, sterile burn		

DISPOSABLES

	Item Used	No. Used	Size
93.	Smock, disposable		
102.	Splint, any size		
61.	Swaddler, silver		
42.	Solution set, mini drip		
41.	Solution set, regular		
113.	Suction, catheter, any		
114.	Suction, cart. disp.		
115.	Suction, tubing		
22.	Suction, V-vac tip		
65.	Suction, yankaeur		
36.	Syringe, 1 cc.		
12.	Syringe, 10 cc.		
37.	Syringe, 60 cc.		
109.	Tape, 1" transpore		
108.	Tape, 2" adhesive		
111.	Thermoscan cover		
48.	Thumper		
107.	Tourniquet, disposable		
99.	Tubex, disposable		
25.	Vacutainer, tube, hldr.		
73.	Venigard		

FLUIDS, MEDICATIONS

	Item Used		
91.	Albuterol		
40.	Alcohol		
16.	Adenocard, 6 mg/2 cc.		
39.	Ammonia inhalant		
53.	Atropine, 1 mg/10 cc.		
66.	Benadryl, 50 mg/1 cc.		
56.	Bretylium, 500 mg/10 cc.		
98.	Charcoal, actidose		
60.	Dextrose, 5%		
64.	Dextrose, 50%		
97.	Diazepam, 10 mg/2 cc.		
67.	Dopamine, 200 mg/5 cc.		
69.	EPI 1:1,000, 1 mg/1 cc.		
70.	EPI 1:10,000, 1 mg/10 cc.		
52.	Hydrogen peroxide		
76.	Isuprel, 1 mg/10 cc.		
72.	Furosemide, 40 mg/4 cc.		
78.	Lidocaine, 100 mg/5 cc.		
79.	Lidocaine, 2 gm/5 cc.		
55.	Mazicon, 5 mg/5 cc.		
84.	Naloxone, .4 mg/1 cc.		
92.	Naloxone, 2 mg/2 cc.		
87.	Nifedipine, 10 mg/2 cc.		
85.	Nitro spray, 4 mg		
86.	Nitrous oxide, prn.		
50.	Normal saline, 1,000 cc.		
63.	So. bicarb., 50 MEQ/50 cc.		
38.	So. chlor. irrigation		
71.	Tetracine, .5%		

HAZARDOUS MATERIALS EXPOSURE FORM

Name:	SSN:
Incident No.:	Date:
Time:	Type of Call:
Location:	
Protection Used:	Supplement Number:

Vital Signs:		
B/P:	Pulse Rate:	Respirations:

Length of Exposure:	Medical Action Taken:
On-Scene Activity:	Material's Hazard Class:
Material's UN ID#:	
Material's Shipping Name:	
Material's Trade Name:	
Material's Bio/Medical Name:	
Other:	

Page 1 of 2

NOTIFICATION OF ACCIDENT REPORT

Note: This report is to be completed in the event of an accident involving a vehicle owned or operated by the department. The report should be accompanied by a copy of the police report, if appropriate. A diagram of the accident scene should be drawn on the back of this form and a narrative of the accident should be attached.

Date of accident:	Day of week:	Time:
Location of accident (including city and state):		
Operator of vehicle:	Station and shift:	Employee ID no.:
Address:		
Home phone no.:	Work phone no.:	Date of birth:
Driver's license no.:	Unit no.:	Vehicle ID no.:
Description of vehicle (including year, make, and model):		
Vehicle license no.:	Was the vehicle responding to an emergency? ☐ Yes ☐ No	
List the names of occupants of the department vehicle and the nature and extent of any injuries:		
Describe the damage to the department's vehicle:		
Name and address of driver of other vehicle(s):		

NOTIFICATION OF ACCIDENT REPORT, *continued*

Name and address of vehicle owner if different from the driver:		
Driver's license no.:	Home telephone no.:	Work telephone no.:
Insured by/policy no.:	Year, make, model, and license no. of other vehicles:	
Names of occupants of other vehicle and the nature and extent of injuries (please indicate where they were transported):		
Describe damage to other vehicle:		
Describe damage to property (fence, light pole, building, etc.):		
Name, address, and telephone number of owner(s) of the property:		
Name, address, and telephone number of witnesses:		
Who was at fault?	What caused the accident?	
Was a citation issued? ☐ Yes ☐ No	To whom?	
Name of investigating officer:	Agency investigating accident:	Report no.

ON-THE-JOB INJURY SUPERVISOR'S REPORT

The boxes on line below are for the use of the Insurer

I.C. file no.	Reason for filing	Insurer code	Insurer location	Insurer claim no.

Employer

1. Name of employer		2. Federal tax I.D. no.		3. Employer's case no. (if applicable)	
4. Mailing address		5. Location (if different from mailing address)			
6. Parent corporation (if applicable)		7. Nature of business			
8. Insurer		9. Policy number		10. Effective date	
11. City or county where accident occurred		12. Employer's premises?		13. State property?	
14. Date of injury	15. Hour of injury	16. Date of incapacity		17. Hour of incapacity	
18. Was employee paid in full for day of injury?		19. Was employee paid in full for day incapacity began?			
20. Date injury or illness reported	21. Person to whom reported	22. Name of other witness		23. If fatal, give date of death	

Employee

24.	Name of employee	25.	Phone number	26.	Gender
27.	Address	28.	Date of birth	29.	Marital status ☐ Single ☐ Divorced ☐ Married ☐ Widowed
		30.	Social security no.		
31.	Occupation at time of injury or illness	32.	Department	33.	No. of dependent children
34.	How long in current job? / 35. How long with current employer?	36.	Was employee paid on a piecework or hourly basis? ☐ Piecework ☐ Hourly		
37.	Hours worked per day / 38. Days worked per week	39.	Value of perquisites per week Food/meals | Lodging | Tips | Other		
40.	Wages per hour / 41. Earnings per week (incl. overtime)				
42.	Machine, tool, or object causing injury or illness	43.	Specify part of machine, etc.		
44.	Were safeguards or safety equipment provided?	45.	Were safeguards or safety equipment used?		
46.	Describe fully how injury or illness occurred				
47.	Describe nature of injury of illness, including parts of body affected				
48.	Physician (name and address)	49.	Hospital (name and address)		
50.	Probable length of disability / 51. Has employee returned to work?	52.	If yes, at what wage	53.	On what date?
54.	EMPLOYER: prepared by (incl. name and title)	55.	Date	56.	Phone number
57.	INSURER: processed by	58.	Date	59.	Phone number

SUPPLEMENTAL REPORT OF INJURY

When and Where to File:

For all injuries that require an On-the-Job Injury Supervisor's Investigation Report to be filed, the employer must file by first-class mail or personal delivery a Supplemental Report of Injury with the employer's workers' compensation carrier and the injured employee: (1) within three days after the injured employee returns to work; (2) within three days when the employee, after returning to work, has an additional day or days of disability because of the injury; (3) within ten days after the end of each pay period in which the employee has had an increase or decrease of benefits; (4) within ten days after the employee resigns or is terminated. If the injured employee is no longer employed by the employer, the employee is responsible for providing information to the carrier about amounts of earnings or offers of employment. The employee may use a Supplemental Report of Injury for this purpose. An employee has disability if he/she is unable to work as a result of the injury or has returned to work earning less than preinjury wages because of the injury.

EMPLOYEE INFORMATION

1. Employee Name: Telephone Number: ( )	2. Social Security No.:	3. Date of Injury:
4. Employee's Mailing Address (Street or P.O. Box)		
City:	State:	Zip Code:

TO EMPLOYER:

Based on the above rule requirements, check the boxes that show the reasons for filing a Supplemental Report of Injury on this date:

☐ Employee returned to work: Complete Blocks 5a or 5b, 6, and 7.

☐ Additional days of disability: Complete Blocks 5b and 7.

☐ Change in weekly earnings after injury: Complete Blocks 7 and 8.

☐ Employee terminated/resigned: Complete Blocks 5a or 5b, 7, and 9.

Supplemental Report of Injury, *continued*

5a. If initial filing, first day of disability due to injury:	5b. If second or subsequent filing, give first day of disability due to injury for this period only:
6. Date of return to work: Check box: ☐ Full duty, full pay ☐ Limited duty, full pay ☐ Reduced pay	7. Weekly and hourly earnings at the time of this report: Weekly: ________ Hourly: ________ Check box: ☐ Same as preinjury wages ☐ Increase from preinjury wages ☐ Decrease from preinjury wages
8. No. of hours working weekly at the time of this report: ________________________ Check box: ☐ Increase from preinjury hours ☐ Same as preinjury hours ☐ Decrease from preinjury hours	9. If the employee resigns or is terminated, fill in the appropriate section: ☐ Date of resignation: ☐ Date of termination:
10. If applicable, eight days of disability began on:	9a. Reason for resignation or termination:
11. Has injured employee died? If so, give date of death:	12. Was employee on limited duty at the time of termination? ☐ Yes ☐ No

EMPLOYER INFORMATION

13. Employer's Business Name:	14. Telephone Number: ( )	
15. Employer's Business Mailing Address (Street or P.O. Box):		
City:	State:	Zip Code:
16. Name of Workers' Compensation Carrier for Above Injury:		

17. The information provided in this report is accurate to the best of my knowledge. It may be relied upon for evaluation of the named employee's eligibility for benefits.

____ ________________

Signature and title of person completing form Date

☐ Employer ☐ Employee

DESIGNATION OF DUTY STATUS FORM

Employee's Name:	Department:	Date:

Check one:
☐ Original injury ☐ Follow-up treatment ☐ Reoccurrence of injury

____
Department head/supervisor

Medical examination reflects the following:

_____ 1. Employee able to perform normal duty.

_____ 2. Employee able to perform light duty (indicate possible activities)

_____ A. Desk job only (sitting only, answering telephones, etc.).

_____ B. Desk/limited movement (some walking and standing, bending).

_____ C. Light physical (walk, stand, lift light objects up to 10 lbs.).

_____ D. Moderate physical (walk, stand, lift objects up to 30 lbs.).

_____ E. Employee can/cannot drive.

_____ F. Other possible activities or limitations: ______________________
____

_____ 3. Employee unable to perform any duties.

_____ 4. Employee's next appointment is: ______________________

_____ 5. Employee referred to ______________________ for further treatment.

_____ 6. Injury is job-related.

_____ 7. Injury is not job-related.

_____ 8. Employee attended physical therapy on: ______________________

Remarks: ______________________________________
____

EMS signature	Physician or therapist signature

Patient: Time in: ______________ Time out: ______________
(Time in/time out to be completed by clinic, doctor, or treatment facility.)

Patient is to return this form to his or her department. Please indicate the nature of the injury, the treatment, and the prognosis.

____

AUTHORIZATION TO RIDE AN APPARATUS RELEASE

I, ________________________________, do hereby release the fire department from any and all responsibility for an accident, injury, or death that may occur while riding on an emergency vehicle. I further understand that I am riding strictly as an observer and will not attempt to take part in any firefighting activity or perform any patient care.

Signed:

Parent or Guardian:

Date:

Approval:

Date:

INSPECTION REPORT FORM

Occupancy			Date
Address			
Mailing Address			
	These premises were found to be reasonably fire safe at this time.		
	Reinspection date		
Occupancy code		Square footage	

Actions Necessary to Correct Hazards Found:

A. Flammable Liquids / Gases / Solids / Explosives

1.		Excessive amount on premises.
2.		Storage rooms—premises not properly marked.
3.		Storage prohibited except in fire-resistive room or outside building.
4.		Valves—dispensing equipment to be approved type and maintained in good order.
5.		Electric wiring and appliances to be UL-approved type.
6.		Discontinue use of any flammable liquid having a flash point below 100° for cleaning.
7.		Post and maintain *No smoking* and *Stop your engine* signs in conspicuous locations.
8.		Provide approved safety cans for all dispensing of flammable liquids.
9.		Securely chain or strap all compressed gas cylinders in a vertical position.
10.		Misuse of materials subject to fire and injury to personnel.
11.		Other hazard.

B. Electric Distribution System

1.		Wiring / switches / plugs defective, to be replaced.
2.		Overload / fixtures / panel not in accordance with N.E.C.

C. Electric Appliances and Equipment

1.		Appliances—motors to be maintained in working order.
2.		Extension cords used as permanent wiring.
3.		Wiring improper / unsafe practice.

D. Fire Protection Systems and Appliances

1.		Check all fire extinguishers for proper maintenance, and recharge all that are expended.
2.		Provide and maintain ___-rated fire extinguishers for each ___ square feet of floor space.
3.		Mount fire extinguishers in conspicuous accessible locations, not more than five feet above floor level.
4.		Standpipe / hose cabinets / valves—repair, replace, obstructed.
5.		Sprinkler system riser, valves, heads, siamese, hydrants, alarms—repair, replace, inoperative, obstructed.
6.		Provide and maintain a minimum of 18" clearance between the top of any storage and sprinkler heads or any overhead obstructions.

E. Heat-Producing Devices		
1.		Defective / to be replaced / to be repaired.
2.		Improper / prohibited device / to be repaired / to be removed.
3.		Venting improper / inadequate.
4.		Combustible stored too close to appliance.
5.		Appliance not approved by A.G.A.
6.		Clearance from appliance inadequate.
7.		Remove accumulation of grease from all cooking apparatus, vent hoods, ducts, vents, etc.
F. Exits, Passageways, Life Safety		
1.		Corridors / passageways obstructed, improperly maintained.
2.		Stairs, towers obstructed, defective, improperly maintained.
3.		Door devices, panic hardware, fusible links defective, inadequate.
4.		Exit passageway inadequately lighted, provide exit lights.
5.		Exit lights to be lighted at all times.
6.		Exit signs—marking inadequate, improper size.
7.		Exits inadequate, insufficient number.
8.		Discontinue practice of locking or blocking any designated exit.
G. General Hazards		
1.		Hazardous accumulation / rubbish / debris / waste materials to be removed.
2.		Smoke and heat detection / fire alarm systems inoperative, to be repaired.
3.		Repair partially burned building or remove it from the premises.
4.		Ventilating systems to be cleaned, repaired, installed.
5.		Aisles, cross aisles to be maintained full width at all times.
6.		Combustible lint and dust to be removed from equipment, walls, beams, floor, and to be properly disposed.
7.		Vacant building—secure all openings until such time as the building is made secure or demolished and removed.
8.		Provide adequate flameproofing for all combustible decorations, drapes, or curtains.
9.		Provide metal containers with metal covers for the collection and storage of waste combustibles.
10.		Provide metal containers with self-closing metal covers for the storage of soiled shop towels and rags.
11.		Replace all missing ceiling tiles.
12.		Post address outside building.
H. Others		
1.		A valid fire department permit required for the following:

Remarks:

MAINTAINING HAZARDOUS CONDITIONS IS IN VIOLATION OF LOCAL ORDINANCES

Inspector: Apparatus and shift:

Copy delivered to:

Type of inspection:

☐ Routine/company insp. ☐ First reinsp. ☐ Second reinsp.

☐ CO ☐ Permit ☐ Special test ☐ Code check

Letters ☐ 1 ☐ 2 Memos ☐ BO ☐ PD ☐ HD ☐ CE ☐ Other

FLASHOVER/BACKDRAFT REPORT

Date:	Time:
Incident No.:	Location:
Name of Person Completing Report:	First-Due Engine Company and Shift:

Description of Structure:

Unusual Contents or Conditions:

Situation on Arrival:

Fire Department Actions:

Sequence of Events:

Warning Indicators:

Injuries/Protective Equipment Experience:

Lesson Learned:

Comments/Observations:

REHAB LOG

Date:	Location:	Incident no.:

Name	Vital signs		Time		Comments
	B/P	Pulse	In	Out	

REHAB UNIT CHECKLIST

Date:		Vehicle I.D. No.:
Odometer:		Onan Hourmeter:
Member Completing Checklist:		

Pass	Fail	Credit card and fuel card in the vehicle: ☐ Yes ☐ No
		Bathroom is clean and there is water for bathroom and sink
		Brakes (service and emergency) function properly
		Cascade system is full
		Doors and compartments open and close properly
		Electrical charging system: Battery A ______ Battery B ______
		Engine starts and runs properly
		Equipment in place and operational
		Food, drinks, and ice
		Fluid levels: Leaks? ☐ Yes ☐ No Added: Oil _____ qts. Gasoline _____ gallons Transmission _____ qts.
		Generator starts and runs properly
		Heater and AC operational
		Lights (headlights, clearance, scene, etc.)
		Microwave oven, refrigerator, sink, coffee pot, etc.
		Oxygen bottle: _____ lbs. Note: Replace if below 500 lbs.
		Steering and suspension
		Tire pressure: LF ___ RF ___ LRI ___ LRO ___ RRI ___ RRO ___
		Warning devices (lights, sirens, backup alarm)
		Wheels and lug bolts
		Windshield washers/wipers

Comments/repairs needed:

EQUIPMENT SERVICE REQUEST FORM

Vehicle/Equipment:	I.D. No:
Date:	Driver:
Description of the Problem:	
Description of the Work/Service Required:	
Authorized By:	Estimated Completion Date:

Copies:	White to Maintenance and Logistics Officer
	Yellow to Station Log

SMALL-VEHICLE WEEKLY CHECKLIST

ID No.:	Date:
Odometer Reading:	Inspector:

Items to Be Checked:	Pass	Fail
Battery Condition		
Brake Fluid Level		
Coolant Level		
Fire Extinguisher		
Fuel Level		
Lights/Turn Signals/Clearance Lights/Horn		
Oil Level		
Radios/Cellular Phones		
Spare Tire Condition		
Tire Pressure: LF____ RF____ LR____ RR____		
Transmission Level		
Upholstery Condition		
Vacuumed		
Warning Devices		
Washed		
Windshield Wipers/Washer Level		

Comments:

Officer's Signature:	Station and Shift:

ANNUAL FIRE PUMP SERVICE TEST FORM

ID No.:	Manufacturer:	
Year Built:	Serial No.:	
Chassis Type/Model:		Serial No.:
Pump:		
Make:	Model:	Type:
Serial No.:		

Item	Start	10 min.	20 min.	30 min.	40 min.
Engine Temp.					
Engine Oil psi					

Engine Hours:			
Pump Capacity:	@ 150 psi:		
	@ 200 psi:		
	@ 250 psi:		
Suction Hose:	Size:	Length:	Lift:
Water Supply Source:			

Test	GPM	Net psi	RPM
150 psi/100% capacity/20 min.			
200 psi/70% capacity/10 min.			
250 psi/50% capacity/10 min.			
Spurt/110% capacity/10 min.			

Comments:

Test Supervisor:	Date:

PREVENTIVE MAINTENANCE WORKSHEET

The maintenance items listed on this worksheet will be performed as indicated by the schedule below:

Quarterly

In-Cab Drive Test:

	1.	Leaks/doors/windows.		**9.**	Parking brakes.
	2.	Seatbelts/air bags.		**10.**	Shifting.
	3.	Wash engine/battery box.		**11.**	RPM shifting.
	4.	Starter operation.		**12.**	Interior lights.
	5.	Fire extinguisher.		**13.**	Legal documents.
	6.	Horn, HVAC, wipers, mirrors.		**14.**	Gauges.
	7.	Steering play: _____ 1.5 max.		**15.**	Floor/wiring under dash.
	8.	Inspection sticker.			

Tires/Wheels:

	16.	Condition: ☐ Good ☐ Fair ☐ Poor		**18.**	Alignment: ☐ OK ☐ Out of Alignment
	17.	Inflation: LF ___ RF ___ LR___ RR ___ S ___ Duals: ________ Tandem: ________		**19.**	Wheels, hubs, lug bolts, etc.
				20.	Rotation (every 10,000 miles):

Exterior:

	21.	Fuel tank caps, lids, etc.		**24.**	Hanging wires, splices, etc.
	22.	Paint, dents, dings, etc.		**25.**	Trunk lid, tailgate doors, side doors on vans, etc.
	23.	Bumpers, reflectors, lenses, mud flaps, etc.		**26.**	Warning devices.

Oil/Lubrication:

	27.	Renew engine oil.		**29.**	Lubricate chassis.
	28.	Renew oil filter.		**30.**	Check all fluid levels.

Underneath Inspection:

	31.	Radiator and engine mounts.		**34.**	Ball joints, hydraulic lines, etc.
	32.	Steering, tie rods, arms, etc.		**35.**	Suspension, springs, shocks, hangers.
	33.	Adjust brakes, check air lines, compressor, etc.		**36.**	Clutch/transmission linkage adjustment.

Underneath Inspection, cont'd:			Preventative Maintenance Worksheet, Page 2 of 2		
	37.	Bell housing.		**40.**	Belts, hoses, wiring.
	38.	Exhaust system..		**41.**	Rear end.
	39.	Driveline.			
Engine Compartment:					
	42.	Battery fluid level.		**46.**	Cables, routing, grounds.
	43.	Load test.		**47.**	Alternator output: ______ amps. ______ volts.
	44.	Clean posts/apply protection.		**48.**	Starter crank amps: __________
	45.	Battery box/hold downs/undercoating.		**49.**	Radiator, hoses, belts, etc.
	46.	Cables, routing, grounds.		**50.**	Fuel lines.

Semiannual

In addition to Items 1 through 50 listed above, the following maintenance items shall be performed on a semiannual basis:

	51.	Wash, compound, and wax vehicle.		**55.**	Lubricate chassis and functional equipment.
	52.	Replace engine oil and filter.		**56.**	Lubricate rods and linkages as required.
	53.	Replenish all fluids as needed.		**57.**	Check and tighten nuts and bolts, including lug nuts on wheels, as needed.
	54.	Replace fuel and air filters.		**58.**	Service auxiliary generator.

Annual

In addition to Items 1 through 58 listed above, the following maintenance items shall be performed on an annual basis:

	51.	Drain transmission, replace filter and fluid.		**54.**	Tune engine.
	52.	Drain rear end, replace fluid.		**55.**	Drain pump transmission, replace fluid.
	53.	Replace hoses and belts as required.		**56.**	Drain aerial hydraulic fluid, replace filter and fluid.

Comments:	
Mechanic:	Date:

Red Tag Out-of-Service Card

Date: ______________________________

Description of the Problem:

____
____
____
____
____
____

Signature: ______________________________

MASTER HOSE RECORD CARD

ID No:	Size:	Manufacturer:	Brand:
Vendor:		P.O. No:	
Type:		Date Received:	
Cost/ft:	Warranty:		Acceptance:
			Test:
			psi:
			Date:

Date:	Tested/psi:	Co. No.:	Remarks:

MONTHLY SCBA INSPECTION FORM

SCBA No.:	Cylinder No.:
Mask No.:	Apparatus:
Date:	Shift:

I. Visual Inspection:

Mask:

	For facepiece-mounted regulators:		**Low-pressure hose.**
	Intermediate pressure hose.		Exhalation value.
	Gaskets.		O-rings, gaskets, and screens.
	Lens, hardware, buckles, straps.		Speaking diaphragm.
	Connections and threads.		Lens, hardware, buckles, etc.
	Connections and threads.		

	Cylinder:		**Regulator:**
	Cylinder pressure.		Values, gauges, and controls.
	Check for gouges, corrosion, chipping, and cracking.		Exhalation valve on facepiece-mounted regulators.
	Gauges and valves.		Gaskets, O-rings, screens, etc.
	Seals, gaskets, and screens.		Alarm.
	High-pressure hose.		High-pressure hose.
	Hydrostatic test date.		

Backpack

	Straps, buckles, cylinder lock, and frame.

II. Operational Inspection

	Don mask, check facepiece seal.		Operate bypass valve.
	Open cylinder value. Check for leaks in hose, alarm or regulator.		Check alarm.
	Operate pressure/demand switch and check for positive pressure.		Clean, recharge, and restore to service. Remove and report any unit that fails to operate properly.

Signature of Inspector:

GROUND LADDER RECORD FORM

Manufacturer's ladder I.D. number or code:	I.D. no.:		
Manufacturer:	Ladder assignment:		
Date of purchase:	Date placed in service:		
TYPE OF GROUND LADDER:	☐ Combination	☐ Extension	☐ Folding
	☐ Roof	☐ Step	☐ Wall
LADDER CONSTRUCTION:	☐ Aluminum	☐ Fiberglass	☐ Wood
	☐ Solid Beam	☐ Truss Beam	
HEAT SENSOR LABEL CHECK:		☐ OK	☐ Failed
Previous repairs, reason for repair, and date of repair:			
Reason and type of test, test date, and person(s) performing test:			
TEST RESULTS:			
HORIZONTAL BENDING TEST:		☐ Pass	☐ Fail
Amount of deformation:			
HARDWARE TEST:		☐ Pass	☐ Fail
ROOF HOOK TEST:		☐ Pass	☐ Fail
HARDNESS TEST:		☐ Pass	☐ Fail
Readings for each test point:			
EDDY CURRENT TEST:		☐ Pass	☐ Fail
Average of three readings and high reading for each test point:			
LIQUID PENETRANT TEST: Location of inspection and results:			
Repairs needed:			
Repairs completed:			
Date completed:			
Person(s) performing repairs:			
Signature:			

TACTICAL WORKSHEET

Incident No.:	**Date:**	**Time:**
Location:		
Occupancy:	**Initial Report by:**	**Time:**

Utilities: ☐ Gas ☐ Electric ☐ Water ☐ Telephone

Systems Support: ☐ Sprinkler System ☐ Standpipe ☐ Fire Pump ☐ Other

Times: ☐ All Clear ☐ Secondary Search ☐ Tap Out

Location of:

Level I Staging:
Level II Staging:
Rehab:
Triage:
Treatment:
Transportation:
Other:

First-Alarm Assignment:

Engines:
Trucks:
Ambulances:
Tankers:
Other:

Second-Alarm Assignment (Use Back for More Alarms):

Engines:
Trucks:
Ambulances:
Tankers:
Other:

Command:

☐ Sector 1:
☐ Sector 2:
☐ Sector 3:
☐ Sector 4:
☐ Water Supply:

☐ Support:
☐ Staging:
☐ EMS:
☐ Other:

Tactical Worksheet, *continued*

Fire Control Evaluation

1. Describe the building/occupancy conditions on arrival and the action taken:

2. Describe the effectiveness of the operation:

3. Describe any special considerations, such as hazardous materials, rescue, welfare efforts, injuries, etc.

4. Describe the salvage and overhaul operations:

5. Describe the condition of the scene for investigators and occupants:

6. Items Requiring Attention:
   - ☐ Procedures.
   - ☐ Apparatus.
   - ☐ Equipment.
   - ☐ Evolutions.
   - ☐ Protective clothing.
   - ☐ Training.
   - ☐ Dispatch response.
   - ☐ Command.
   - ☐ General operations.

7. What operations would you change?

8. What operations worked well? Why?

9. Evaluator:

10. Date:

11. Shift:

MINIMUM COMPANY STANDARDS EVALUATION FORM

Company Tested:	Date:

Location of Tests:

Instructions:
As per SOP 600.03, each company shall successfully perform one EMS evolution, two fire evolutions, and one driver evolution on an annual basis.

EMS Evolution	Time Allocated (Minutes)	Time Required
#1 MAST application	05:00	
#2 Traction splint	05:00	
#3 KED(TM) application	08:00	
#4 Trauma assessment	10:00	
#5 Thumper	08:00	
Pass:	**Fail:**	**Score:**

Participants:

1.	5.
2.	6.
3.	7.
4.	8.

Comments:

Rater:	Date:

Minimum Company Standards Evaluation Form

Fire Evolution	Time Allocated (Minutes)	Time Required
1. SCBA	01:00	
2. Ladder raise	14 ft.: 00:30	
	24 ft.: 01:30	
	35 ft.: 02:00	
3. Lights/fan	03:00	
4. Truck operations	05:00	
5. 1 3/4-inch handlines	04:00 (4)	
	04:30 (3)	
6. Two-piece ops	05:30 (6)	
	06:00 (8)	
7. Deck gun	03:00	
8. Elevated M.S.	05:00	
9. A-S support	04:00	
10. Portable deluge	04:30 (4)	
	05:00 (3)	
11. Knots and hitches	Not timed	Pass ☐ Fail ☐
Pass:	**Fail:**	**Score:**

Participants:

1.	5.
2.	6.
3.	7.
4.	8.

Comments:

Rater:	**Date:**

Minimum Company Standards Evaluation Form

Driver Evolution	Pass	Fail
1. Serpentine		
2. Alley dock		
3. Opposite alley		
4. Diminishing clearance		
5. Straight line		
6. Turning around		

Person Being Evaluated:

Date of Evaluation:

Weather Conditions:

Comments:

Rater:

TACTICAL SURVEY FORM

Building being inspected		Date	Copy for (circle): Company Battalion Bureau
Name of owner, occupant, lessee, tenant		Inspector	
Address	Phone	Company	Platoon / Battalion / Division
Name of responsible party who can be reached at night		Make one diagram for the complex and one diagram for each fire area or building.	
Address	Phone	This is page _____ of a set of _____ pages.	

SHOW ON EVERY DIAGRAM:

1. Arrow north
2. Street and number
3. Height
4. Fire walls
5. Stairs and fire escapes
6. Doors
7. Vertical openings
8. Scale (1" = 10', 25', or 50', etc.)

Street name and address

Name of company or occupancy

HAZARD

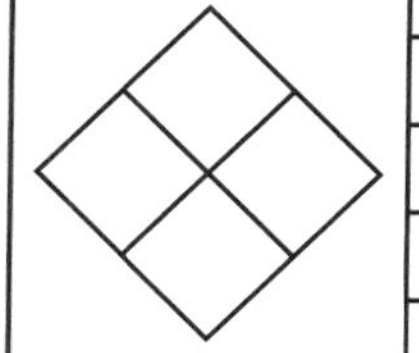

Responding companies:	Engine	
Trucks	Squads	Snorkels
Arterial	Water main	Cistern
	Size	
Street		

Circle as appropriate:

1. **SECURITY:** Guards, fenced, locked. Dogs. Fire alarm, central watch, ADT. DO NOT force entry through: transom, side door. LADDER: front, rear, left, right, other.
2. **SECOND EXIT?** Explain how to get in: ______________________________
3. **OCCUPANCY:** Church, school, institution, hospital, offices, nursing home, apartment house, residence, factory, garage, hotel. STORES: Retail, wholesale, warehouse, other.
4. **BUILDING:** Height _____ and basement area _____ × _____. Year built: ________. **CONSTRUCTION:** Mostly, all: Frame, metal-clad, mill, ordinary, fire-resistive, all metal.
5. **WALLS:** Wood, metal-clad, concrete block, brick, veneer. WINDOWS: None, few, covered, ordinary, other, storefront, wireglass, factory type, glass block, barred.
6. **ROOF:** Safe, not safe. Scuttle, doors, skylights, hip, flat, truss, slate, tile, composition, gravel. Metal on: steel, wood rafters, bar joist. Plywood.
7. **ROOF ATTACHMENTS:** NONE, penthouse, gravity tank, cornice, ventilators, coolers, air conditioners, signs, canopy. CEILING: ____________________. Remodeled.
8. **ADDITIONS:** NONE, passageway, docks, porches, sheds (# of ____ trucks), garage, other: ________________________. CONDITION: good, fair, poor.
9. **HYDRANTS:** ______-inch main. Fed ______ ways. Flushed. Drains, doesn't drain. Threads tested. Needs: Repairs, cap(s). Distance to this building: __________ feet.
10. **SPRINKLERS:** NONE, not supervised. All: Wet, dry. All not heated, possible freeze, open windows. Unheated: dock, penthouse, cooler. DPV: Antifreeze loop.
11. **SECTIONS:** (All, some) NOT SPRINKLERED. (All, some) SECTION VALVES: Open, closed, not marked. ______ ft to top sprinkler head. Single, several sources of water supply.
12. **WATERFLOW ALARMS:** NONE, electric, water motor gong. Inside, outside of building. NOT TESTED. CONDITION: good, fair, poor. PIPING: Rusty, loose, missing, sagging.
13. **HEADS:** Painted, bent, old, corroded, dirty. Obstructed by: curtains, drop ceiling, partitions, tables, stock. EXTRA HEADS: Ample, few, NONE, no wrench.
14. **DPV:** NONE, AIR PRESSURE: (#_____ riser _____ lbs) (#_____ riser _____ lbs) (#_____ riser _____lbs) WATER PRESSURE: _____ lbs. GAUGES: Missing, broken.
15. **GRAVITY TANK:** NONE, on roof, in tower. Empty, 1/2, 1/2, 3/4, full. Frozen, leaking. SUPPORTS: Rusty, weak. TANK VALVE: Open, closed.
16. **PRESSURE TANK:** NONE, in basement, in tower. CAPACITY: _____,000 gals. 2/3 full, down _____ inches. AIR PRESSURE: _____ lbs. TANK VALVE: Open, closed.
17. **FIRE PUMP:** NONE, OFF, CAPACITY: _____ gpm. On: Manual, automatic. MAINTAINS: _____ lbs pressure. VALVE(S): Open, closed. CONDITION: good, fair, poor.
18. **SIAMESE:** NONE, not accessible, not hidden. _____ feet to hydrant. THREADS: N.S. or FD, not damaged, not free, not capped, no gasket, no clapper, no sign.
19. **STANDPIPE:** NONE, _____ lbs pressure at base. TAKEOFF VALVES: 2 1/2 and 1 1/2, accessible, blocked. THREADS: N.S. or FD, no handles, not capped.
20. **NO. OF RISERS:** _____ , in stairwells, end, center of corridors, center of floor. HOUSE LINE: NONE, racked (yes no), reeled (yes no), missing, old, no nozzle. CONDITION: good, fair, poor.

SPRINKLERS

21. **FIRE ALARM BOX:** NONE, _____ feet from entrance, _____ blocks from entrance. Box #_____ at _________________ Street. Not connected to inside alarm system.

22. **FIRE ALARM SYSTEM:** NONE, no. of heat-activated detectors: _____; smoke detectors: _____ CODE: Posted, not posted. System is: ON, OFF. Control panel: Location, keys (where?) (reset how?).

23. **EXTINGUISHERS:** NONE, insufficient, not on brackets, empty, not tagged, dated __________. NEED: Several, few. Class: A, B, C, D, repairs.

24. **FIRE DOORS:** NONE, _____ automatic, _____ self-closing. One, more than one: damaged, blocked, vented, missing, not labeled, inoperable.

25. **HEATING BY:** (Hot-water, low, medium, steam) boiler. Furnace, stove(s), unit heater(s). Coal, oil, gas. No combustibles too close. CONDITION: good, fair, poor.

26. **SMOKEPIPE:** Not safe. Clearance to combustibles: _____ inches. CONDITION: good, fair, poor. CHIMNEY: Dangerous, cracked, leaks, leaning. CONDITION: good, fair, poor.

27. **ELECTRICAL EQUIPMENT:** CONDITION: good, fair, poor. WIRING: Loose, temporary, bare, open, frayed. HIGH VOLTAGE. MOTORS: Dirty, oily. CONDITION: good, fair, poor.

28. **CIRCUITS:** Not overfused, circuit breakers. Fixtures, light bulbs: touching, too close to: combustibles, cloth, string, paper. ELECTRICS (MAIN) SWITCH: Where?

29. **GAS METERS:** NONE, basement, 1st floor. _____ feet in from: N E S W wall. Near, on, N E S W wall, center of _______________________.

30. **SHUTOFF VALVE:** (Bldg. service pipe) Accessible, inaccessible, paved over, is located _____ feet out from N E S W wall and _____ feet in from N E S W corner.

31. **STOCK:** NONE, very damageable from: smoke, water. DANGEROUS. HIGHLY COMBUSTIBLE: _______________________. Many, few covers needed.

32. **STOCKPILING:** NONE, not straight, tipping, overloaded, too high, not on skids. Loose. AISLES: NONE, blocked, narrow, haphazard, needed.

33. **FLOOR:** Unsafe, overloaded, crowded, littered. Unprotected: pit, openings, holes. FLOOR DRAIN: Not marked, sump pump. How? _______________________

34. **RUBBISH:** NONE, some, much, extreme. Trash, junk, boxes, cartons, baskets, rags, bldg. debris. OTHER _______________.
HOUSEKEEPING: All, in part: good fair poor.

35. **SPECIAL HAZARDS:** NONE. DANGEROUS: Railings, floors, porches, stairs, walls, roof, fire escape. ACCESS BLOCKED: Inside, outside. PRESSURIZED GASES.

36. **HIGHLY VOLATILE:** Solids, liquids, chemicals. Gasoline in: (ordinary, safety) cans, drums. TANK. FUEL OIL: Nonsiphon; (Inside, outside) fill, vent.

37. Flammable liquid vault, dip tank, spray booth, oven, deep fryer, tempering tank, acid carboys, welding equipment. Plastic. Foam rubber, incinerator.

38. **RESCUE:** _____ children, _____ blind, _____ deaf, _____ bedridden, _____ aged, _____ crippled, _____ detained, ____________________ other.

DIRECTIVE TO PHYSICIANS

For Persons 18 Years of Age and Over

DIRECTIVE made this _______ day of ___________________________ (month, year).

I, ____, being of sound mind, willfully and voluntarily make known my desire that my life shall not be artificially prolonged under the circumstances set forth below, and do hereby declare:

1. If at any time I should have an incurable or irreversible condition caused by injury, disease, or illness certified to be a terminal condition by two physicians, and where the application of life-sustaining procedures would serve only to artificially prolong the moment of my death and where my attending physician determines that my death is imminent or will result within a relatively short time without application of life-sustaining procedures, I direct that such procedures be withheld or withdrawn, and that I be permitted to die naturally.
2. In the absence of my ability to give directions regarding the use of such life-sustaining procedures, it is my intention that this DIRECTIVE be honored by my family and physicians as the final expression of my legal right to refuse medical or surgical treatment and accept the consequences from such refusal.
3. If I have been diagnosed as pregnant and that diagnosis is known to my physician, this DIRECTIVE shall have no force or effect during the course of my pregnancy.
4. This DIRECTIVE shall be in effect until it is revoked.
5. I understand the full import of this DIRECTIVE, and I am emotionally and mentally competent to make this DIRECTIVE.
6. I understand that I may revoke this DIRECTIVE at any time.
7. I understand that Texas law allows me to designate another person to make a treatment decision for me if I should become comatose, incompetent, or otherwise mentally or physically incapable of communication. I hereby designate ________________, who resides at ___________________________________, to make such a treatment decision for me if I should become incapable of communicating with my physician. If the person I have named above is unable to act on my behalf, I authorize the following person to do so:

   Name: ____

   Address: ____

I have discussed my wishes with these persons and trust their judgment.

(NOTE: This clause is optional. You do not have to designate another person to make treatment decisions.)

8. I understand that if I become incapable of communication, my physician will comply with this DIRECTIVE unless I have designated another peson to make a treatment decision for me, or unless my physician believes this DIRECTIVE no longer reflects my wishes.

   Signed ____

   City, county, and state of residence ____

Two witnesses must sign the DIRECTIVE in the spaces provided below.

I am not related to the declarant by blood or marriage; nor would I be entitled to any portion of the declarant's estate on his deceazse; nor am I the attending physician of the declarant or an employee of the attending physician; nor am I a patient in the health care facility in which the declarant is a patient, or any person who has a claim against any portion of the estate of the declarant upon his decease. Furthermore, if I am an employee of a health facility in which the declarant is a patient, I am not involved in providing direcvt patient care to the declarant, nor am I directly involved in the financial affairs of the health facility.

Witness ____

Witness ____

DIRECTIVE TO PHYSICIANS

For Persons Under 18 Years of Age

DIRECTIVE made this ________ day of ________________________ (month, year).
On behalf of _________________________, a qualified patient under the Texas Natural Death Act who is under 18 years of age, I/we, ______________________, being of sound mind, willfully and voluntarily make known my/our desire that his/her life not be artificially prolonged under the circumstances set forth below, and do hereby declare:

1. If at any time the patient whose name appears above has an incurable or irreversible condition caused by injury, disease, or illness certified to be a terminal condition by two physicians, and where the application of life-sustaining procedures would serve only to artifically prolong the moment of his/her death and where his/her attending physician determines that his/her death is imminent or will result within a relatively short time without application of life-sustaining procedures, I/we direct that such procedures be withheld or withdrawn, and that he/she be permitted to die naturally.
2. On behalf of the said patient, it is my/our intention that this DIRECTIVE shally be honored by his/her physicians as the final expression of my/our legal right to refuse medical or surgical treatment on behalf of the said patient and to accept the consequences from such refusal.
3. If she has been diagnosed as pregnant and that diagnosis is known to her physician, the DIRECTIVE shall have no force or effect during the course of her pregnancy.
4. This DIRECTIVE shall be in effect until it is revoked. I/we understnad that my/our authority to execute this DIRECTIVE on behalf of the above-named patient expires on his/her 18th birthday.
5. I/we understand the full import of this DIRECTIVE and I/we am/are emotionally and mentally competent to make this DIRECTIVE.
6. I/we understand that the desire of the above-named patient, if mentally competent, to receive life-sustaining treatment shall at all times supersede the effect of this DIRECTIVE.

Signed ____

City, county, and state of residence ____________________________________

Indicate relationship to patient: ______ spouse ______ parents _______ legal guardian

Two witnesses must sign this DIRECTIVE in the spaces below.

I am not related to the declarant by blood or marriage; nor would I be entitled to any portion of the declarant's estate on his deceazse; nor am I the attending physician of the declarant or an employee of the attending physician; nor am I a patient in the health care facility in which the declarant is a patient, or any person who has a claim against any portion of the estate of the declarant upon his decease. Furthermore, if I am an employee of a health facility in which the declarant is a patient, I am not involved in providing direcvt patient care to the declarant, nor am I directly involved in the financial affairs of the health facility.

Witness ___________________________________

Witness ___________________________________

VA

Name

Age

Injuries

Treatment

Immediate 1st Priority

Name
Ambulance
Agency ______ No. ______

City/County ______

Destination ______

VA

Immediate 1st Priority

(Front)

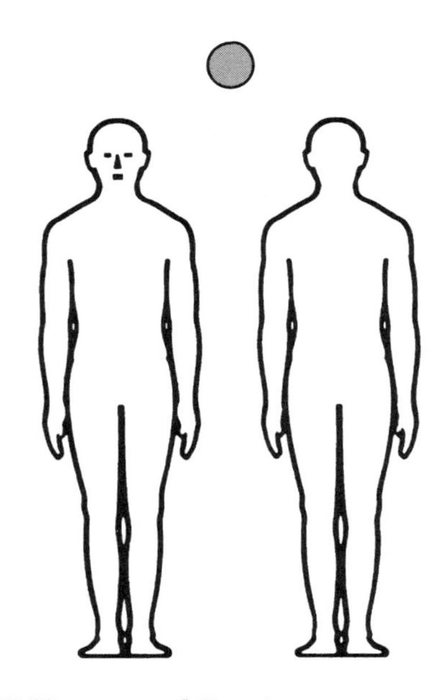

- ☐ Uncorrected Respiratory
- ☐ Problems
- ☐ Cardiac Arrest
- ☐ Severe Blood Loss
- ☐ Unconscious
- ☐ Severe Shock
- ☐ Open Chest or Abdominal Wounds
- ☐ Burns Involving Respiratory Tract
- ☐ Several Major Fractures

(Back)

VA

Name

Age

Injuries

Treatment

Secondary 2nd Priority

Name
Ambulance
Agency ____________ No. ______

City/County ____________

Destination ____________

VA

Secondary 2nd Priority

(Front)

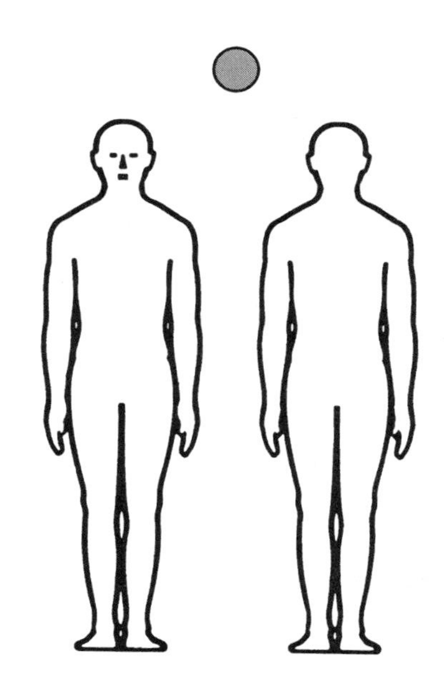

☐ Severe Burns

☐ Spinal Column Injuries

☐ Moderate Blood Loss

☐ Conscious with Head Injuries

(Back)

VA

Name

Age

Injuries

Treatment

Delayed **3rd Priority**

Name ______
Ambulance Agency ______ No. ______
City/County ______
Destination ______

VA

Delayed **3rd Priority**

(Front)

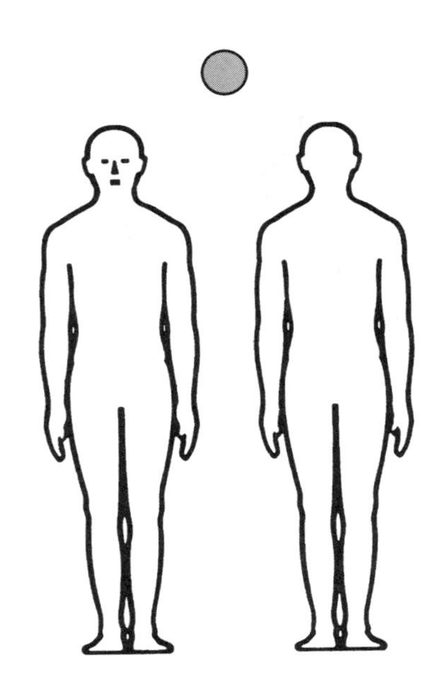

☐ Minor Fractures

☐ Contusions—Abrasions

☐ Minor Burns

(Back)

VA

Name

Age

Injuries

Treatment

Name ______
Ambulance Agency ______ No. ______

City/County ______

Destination ______

VA

Deceased 4th Priority

(Front)

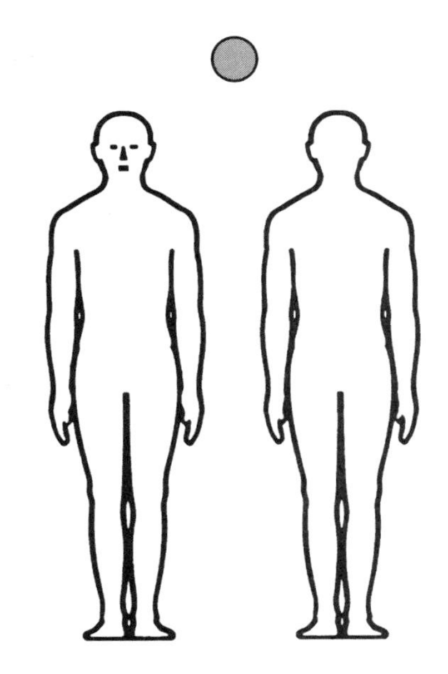

(Back)

EMPLOYEE IMMUNIZATION REFUSAL FORM

I do hereby acknowledge that I was offered the immunization(s) listed below free of charge by my employer and of my own free will declined to accept the immunization(s). I further acknowledge that the immunization(s) were offered to me by my employer as a preventive measure and as a part of the department's Occupational Safety and Health Program.

List immunizations (use back if necessary):

Name of Agency:

Employee Name (Print or Type):

Employee I.D. No.:

Employee Signature:

Date:

ALARM AND FIRE RECORD CARD

Alarm No. __________ Address

Date: __________ Box: __________ Map: __________ Callback No.:

Nature of Alarm: Structure / Grass / Car / Trash / Ambulance Assist / Other

ALARM	DISPATCH	EN ROUTE		ON SCENE		
		In Service		Tapout En route hospital		
		Additional		Out at hospital		

Temp: ________ Wind: ________ Humidity: ________%

Units responding: ______________________________

Alarm source: ____________________ Census: __________

Shift: ______________ Alarm Operator: __________________

REINSPECTION FEE SCHEDULE

Occupancy Size in Square Feet	First Reinspection	Second Reinspection
≥ 3,000	$ 15.00	$ 30.00
3,001 to 6,000	$ 25.00	$ 50.00
6,001 to 12,000	$ 35.00	$ 70.00
12,001 to 18,000	$ 45.00	$ 90.00
18,001 to 24,000	$ 60.00	$120.00
24,001 to 100,000	$100.00	$200.00
≥ 100,000	$200.00	$400.00